D0983123

River and channel revetments

A design manual

Manuela Escarameia

Published by Thomas Telford Publications, Thomas Telford Ltd.
1 Heron Quay, London E14 4JD

URL: http://www.t-telford.co.uk

First published 1998

Distributors for Thomas Telford books are
USA: ASCE Press, 1801 Alexander Bell Drive, Reston, VA 20191-4400
Japan: Mauruzen Co. Ltd, Book Department, 3–10 Nihonbashi 2-chome, Chuo-ku, Tokyo 103
Australia: DA Books and Journals, 648 Whitehorse Road, Mitcham 3132, Victoria

Cover picture shows the River Thames at Wallingford, UK

A catalogue record for this book is available from the British Library

ISBN 0 7277 2691 9

This book corresponds to Environment Agency R&D Publication 16. The work was funded as part of the Environment Agency's National R&D Programme under Project W5-029.

Typeset by Gray Publishing, Tunbridge Wells, Kent
Printed and bound in Great Britain by Redwood Books, Trowbridge, Wiltshire

Preface

The compilation of information on river bed and bank revetment systems, as well as the updating of the existing knowledge with recently developed formulae, was seen by the UK Department of the Environment, Transport and the Regions (DETR) as filling a gap in the current technical literature. This book is the result of the research project that was sponsored by the DETR with additional funding from HR Wallingford and the UK Environment Agency. The main objective is to make the available information easily accessible to design engineers. To achieve this, a large number of photographs, diagrams and worked examples are included; also, simple procedures for each stage of the design process are proposed and data on revetment systems are compiled in the form of Data Sheets. Although efforts were made to produce a comprehensive list of the revetment systems available to UK engineers, the list produced may not be exhaustive and for any omissions I apologise.

By choosing the title 'River and channel revetments' for a book on *engineered* systems, I hope I will not be disappointing those who advocate the sole use of natural materials. Recently a trend has been observed in the UK that recommends the use of vegetation and fibre mats as the sole means of bank protection even in situations where the severity of the current and/or wave attack demands the use of stronger materials such as riprap or concrete. Cases of inadequate design result in expensive repair works and do nothing to enhance the reputation of vegetation as the reliable bank protection material that it can be. Among other aims, this book is intended to give substantiated guidance on the limits of application of revetments, which I hope will contribute to more accurate and economical design.

Frequent reference is made in the text to two other publications: the earlier book by Hemphill and Bramley (1989) from which much information was retrieved, and a publication by Morgan *et al.* (1998), which was produced concurrently with this book. These publications do not give specific guidance on bed protection or cover situations of high turbulence in the flow, but both provide very valuable complementary information on the adoption of management strategies to reduce the risk of bank instability and the use of vegetation for protection of banks against erosion.

This book was conceived as a design manual to provide river engineers with a user-friendly publication that comprises much of the information they require for design. It is appreciated that readers will have varying backgrounds, and while some may benefit from the basic hydraulics and geotechnical information presented here, others will, I trust, find this book a useful aid to design and a concise source of data on river revetment types.

M. Escarameia

Manuela Escarameia *graduated in Civil Engineering (Structures) from the Technical University of Lisbon, Portugal, and obtained a Master of Science degree in Hydraulic Engineering from the University of Newcastle upon Tyne. Before joining Hydraulics Research Wallingford in 1991, she worked in LNEC (National Laboratory of Civil Engineering, Lisbon) in various studies of hydraulic structures. She is currently part of the Water Management Group of HR Wallingford, where she has worked in research and consultancy projects in urban drainage and river engineering, with particular emphasis on protection of river banks and bed.*

Acknowledgements

The project from which this book derives was funded by the Construction Directorate of the Department of the Environment, Transport and the Regions (DETR) with contributions from HR Wallingford and the Environment Agency (EA).

A number of individuals, organisations and companies contributed to this book. The helpful guidance of the steering group that was formed for this project is particularly acknowledged. The steering group was formed by:

A. B. Ibbotson (Ministry of Agriculture Fisheries and Food)
R. W. P. May (HR Wallingford)
J. F. O'Hara (Independent Consultant, formerly of Costain International)
A. T. Pepper (EA Flood Defence R&D Topic Advisor)
R. M. Young (Binnie Black & Veatch)

Acknowledgement is also due for the technical advice, assistance and data kindly provided by:

J. Ackers (Binnie, Black & Veatch)
J. Coates, S. Pedder and A. Sansom (EA North-East Region)
R. Bennet, S. Crowe and H. Yates (EA North-West Region)
G. Tustin (EA Midlands Region)
S. Peck (EA Anglian Region)
M. Luker and D. Murphy (EA Thames Region)
B. Hornigold (Internal Drainage Board)
N. W. H. Allsop and K. J. McConnell (HR Wallingford)
D. W. Knight (University of Birmingham)
A. J. Collins (Silsoe College, Cranfield University)
J. R. Dale (Posford Duvivier, Underwater Division)
S. Dunthorne and R. Wingfield (Sir Alexander Gibb)
A. Pinkett (Independent Consultant)

The help of the following manufacturers/suppliers of river revetments in providing technical data and illustrative material is gratefully acknowledged:

Maccaferri Ltd
MMG Civil Engineering Systems Ltd
ABG Limited
Comtec (UK) Ltd
Cooper Clarke Group PLC
Grass Concrete Ltd

Hesselberg Hydro 1991 Ltd
Intrucel
Proserve Limited
Tensar-Netlon Ltd
Ruthin Precast Concrete
Weldmesh Land Reinforcement/Tinsley Wire

HR Wallingford is an independent specialist research, consultancy, software and training organisation that has been serving the water and civil engineering industries worldwide for over 50 years in more than 60 countries. HR Wallingford aims to provide appropriate solutions for engineers and managers working in:

- water resources
- irrigation
- groundwater
- urban drainage
- rivers
- tidal waters
- ports and harbours
- coastal waters
- offshore.

Address: Howbery Park, Wallingford, Oxon, OX10 8BA, UK
Internet: http://www.hrwallingford.co.uk

Notation

A	cross-sectional area of the flow
B	channel width
C_s	stability coefficient in Equation (22)
C_T	blanket thickness coefficient in Equation (22)
C_v	velocity distribution coefficient in Equation (22)
C	turbulence coefficient, coefficient in Equation (46)
c'	effective cohesion
D	size of protection
D_{n50}	stone size
D_o	coefficient in Equation (8)
D_p	diameter of ship propeller
D_x	size of particle for which x% of the sample in weight is smaller
d_s	scour depth
e	voids ratio, defined as the ratio of the volume of voids to the volume of solids
F	fetch
f	friction
G	coefficient in Equation (14)
g	acceleration due to gravity
H	wave height
H_i	maximum wave height
H_s	significant wave height
h	wave run-up
I_P	plasticity index
I_r	Iribarren number
K_h	depth factor in Equation (19)
K_s	slope factor in Equation (19)
K_T	turbulence factor in Equation (19)
k_d	side slope term
k_1	side slope correction factor in Equation (22)
k_g	permeability of geotextile
k_s	roughness height, permeability of the soil
L	displacement length
L_w	length
n	porosity, defined as the ratio of the volume of voids to the total volume
O_x	opening size of geotextile
P_d	ship engine power
Q	flow
R	hydraulic radius, centreline radius of a bend
TI	turbulence intensity

T_z	wave period
t	thickness of Open Stone Asphalt (OSA)
S	energy slope
s	relative density of stone, defined as $s = \rho_s/\rho$
U	mean cross-sectional velocity
U_b	bed (or bottom) velocity
U_d	depth-averaged velocity
U_{10}	wind speed at a height of 10 m above mean water level
u	porewater pressure
y	water depth
W	width of water surface, weight of particle
W_n	component of particle weight normal to the slope
W_s	component of particle weight in the direction of the slope
W_x	weight of particle for which $x\%$ of the total sample is lighter
Z_b	vertical height from the boundary to the propeller axis of a ship
α	angle of bank slope to the horizontal
β	angle of interference peaks of secondary boat waves
ρ	water density
ρ_s	density of stone
σ'_n	effective normal stress
τ	effective shear strength
τ_0	shear stress
ϕ	angle of repose of the soil, stability correction factor in Equation (19)
ψ	stability factor
Ω	coefficient for reduced stability of revetments on banks

Glossary

Anchor	Device to fix a revetment or piled wall into the ground
Apron	Layer of stone, concrete or other material used to protect structures or banks against scour
Bioengineering	The sole use of vegetation for protection against erosion (also known as soft engineering)
Biotechnical engineering	A combination of vegetation and structural units for bank erosion protection
Block revetments	Revetment systems formed by pre-cast concrete blocks, either loose or connected by cables or by an underlying geotextile
Block stone	Large, cuboid stone units, typically heavier than 1000 kg used for bank protection
Boat wash	Term used to describe water movement due to boat motion
Box gabions	Cube-shaped containers usually made of wire or polymer mesh and filled with stone
Cohesion	The resistance due to the forces tending to hold the particles together in a soil mass
Conceptual design	Design stage concerned with the evaluation of the erosion problem, selection of the strategy to control erosion and choice of suitable kind of revetment
Cover layer	Outer layer of a revetment system used for protection against external hydraulic loads
Detailed engineering	Design stage that involves the detailed design and specification of the engineering revetment systems, including filters
Drawdown	Rapid drop in water level, which may lead to excessive hydrostatic pressures behind a revetment layer
Fetch	The length of water body over which the wind can blow to generate waves at a particular point
Filter	Intermediate granular layer or geotextile used to prevent the fine grains of the underlayer from being washed through the larger voids of the cover layer
Flexible forms	Category of revetment that includes sacks and mattresses made of natural or artificial fibres and filled with concrete or sand
Formation	The soil base on which the revetment is to be built or installed

Gabions	Generic name given to revetments consisting of stone contained in steel or polymer mesh. Types include box gabions, gabion mattresses and sack gabions
Gabion mattresses	Rectangular-shaped containers made of wire or polymer mesh and usually filled with stone
Geomat	Three-dimensional geotextile used to increase bank stability
Geotextile	Synthetic, permeable textile, mesh or net used as filter or separation layer
Grouting	Way of improving stability of revetments by filling joints or gaps with cement or bitumen mortars
Hand pitched stone	Revetment usually formed by single sized stone placed by hand in a single layer
Internal friction	The resistance due to interlocking of the particles
Outline design	Design stage that involves comparison of various types of revetment leading to the choice of revetment
Pore pressure	The pressure of air or water in the interstices between particles of soil, rock or other materials
Revetment	A system with no slope retention capability, typically formed by a cover layer and a filter, used to protect sloping banks and beds against erosion
Rhizome	An underground stem with both roots and shoots
Riprap	Randomly placed loose quarry stone used for protection against erosion
Sack gabions	Tubular containers made of wire or polymer mesh and filled with stone
Scour	Local removal of soil particles by hydraulic forces
Screw-race	High-velocity water jet caused by the movement of boats in a waterway
Seepage	The movement of water through the interstices of the river or channel bed or banks (either from the water body to the bed or banks or vice versa)
Shear strength	Maximum resistance to shearing stresses
Shear stress	The drag force exerted by flowing water on the wetted perimeter of a river or channel (defined as a force per unit area)
Significant wave height	The average height of the highest one third of the waves in a given period
Soil reinforcement system	Cellular system formed by strips of polymer or geotextile material used for soil confinement
Structural engineering	The sole use of hard units (made of concrete, rock, steel, etc.) for erosion protection
Toe	The base of a river or channel bank
Turbulence	Random fluctuation of flow velocity around the mean value

Turbulence intensity	Ratio of the variation of flow velocity around the mean and the mean flow velocity near the bed
Underlayer	The layer underneath the cover layer that makes the transition to the underlying soil; it may consist of a granular material or a geotextile
Up-lift forces	Upward forces in the interstices of a particulate material or underneath a revetment layer
Watercourse	Flowing body of water, either natural (river) or built (channel)
Waterway	Navigable watercourse
Wave run-up	The upward movement of waves onto a bank or wall

Contents

Illustrations

Tables

Figures

Introduction

1. Introduction

1.1. BACKGROUND

For its high scenic, agricultural or urban value, the land adjacent to rivers has long been a primary concern of hydraulic engineers in particular and of the population in general. The preservation of rivers and channel boundaries is a major issue involving a wide range of disciplines that deserves an integrated approach if the most appropriate solutions are to be implemented. This book gives guidance on the hydraulic design of such solutions and highlights the need for consideration of other aspects (such as maintenance procedures), but does not attempt to cover these in detail.

Some very good publications are currently available dealing with the subject of protection of rivers and channels against the erosive action of the flow. However, although they provide comprehensive information in many areas, it was thought that there were still some important issues that ought to be addressed.

The first task was to present the information to the river engineer in the form of a design manual rather than in textbook format, as was the case with previous publications. The objective was to structure the information contained in the book in such a way as to make it more easily accessible to engineers of various degrees of experience and knowledge.

Recent research on certain topics not fully covered by the existing literature (such as stability of revetments in highly turbulent environments) also dictated the need for a book that would include new design formulae. Similarly, independent guidance was lacking on the choice between several kinds of revetment in commercial use in the UK and elsewhere. Therefore, a compilation of the major types of river revetment system was considered useful.

The production of this book coincided with the preparation of another publication funded primarily by the UK Environment Agency (EA): *Waterway bank protection: a guide to erosion assessment and management* (Morgan *et al.*, 1998). Although completely separate from each other, the two publications have complementary objectives and therefore the reader of the present book is frequently directed to the above publication. Its scope involves the two types of bank protection system (revetments and vertical walls) but bed protection is not considered; the systems described range from bioengineering to structural solutions (see Section 1.2 for definitions), with particular emphasis on UK practices and environmental conditions. The EA manual gives guidance on the selection of suitable bank protection methods in an essentially qualitative way, whereas this book provides quantitative information for the design of appropriate revetments. The scope of this book is detailed in the following section.

Another manual has been produced by HR Wallingford addressing the design of coastal revetments, which are particularly exposed to wave attack (see McConnell, 1998). The reader should refer to this or to similar publications for the design of bank protection in situations of severe wave attack (e.g. estuaries or large reservoirs).

This book was funded primarily by the UK Department of the Environment (currently Department of the Environment, Transport and the Regions) with contributions from the Environment Agency, Ministry of Agriculture, Fisheries and Food, Binnie Black and Veatch, Costain and HR Wallingford Ltd. A steering group was formed with representatives of the above contributing organisations so that the book would benefit from views of experts in the private and public sectors.

1.2. SCOPE

This book gives guidance and formulae for the design of revetments used to protect banks and beds of rivers and channels against flow-induced erosion. Dealing solely with revetments, the book does not cover in detail systems used for slope retention such as gravity walls. However, because of their extensive use in river works, some information is given on certain types of vertical bank protection (for example, piling).

The types of revetment that are covered in this book are in some publications termed 'engineered' or 'hard' (as opposed to 'soft' revetments consisting solely of vegetation). However, these terms have not been adopted here. For consistency with recent publications the terms used are:

- *bioengineering* – corresponding to the traditionally termed soft revetments, which are outside the scope of this book
- *biotechnical revetments* – those revetments that incorporate some form of vegetative protection but also rely on the technical ability of harder materials (typical examples are grassed concrete blocks)
- *structural revetments* – revetments formed exclusively by non-live materials (examples include concrete lining and riprap).

The types of revetment covered in this book are:

- biotechnical revetments
- structural revetments.

References are also made, with illustrative examples, to revetments formed by more than one type of protection (composite revetments), which reflect the increasing concern with promoting ecologically varied environments. This book is intended to provide information on revetments for application worldwide, but specific reference is made to proprietary systems available in the UK and, in some cases, to design conditions existing in the UK. When confronted by conditions very different from those in the UK, the engineer is obviously urged to seek local design data and information on traditional and well-proven construction practices.

The hydraulic design conditions covered are:

- river and channel current flow (velocities up to 7 m/s)
- waves, both wind- and ship-induced (wave heights up to 1 m)
- flow in bends
- tidal aspects
- high flow turbulence (excluding hydraulic jumps).

The above conditions exclude situations of severe wave attack as encountered in estuaries and in large bodies of water such as reservoirs (see, for example, McConnell, 1998).

The situations in which the measures detailed in this book may be applied are therefore:

- natural rivers and artificial channels
- navigable waterways
- vicinity of hydraulic structures
- tidal reaches.

Limited information is also given on the protection of river beds around bridge piers and other similar structures.

1.3. USE OF THIS BOOK

The book is structured so that the design procedure is given in Chapter 3, and particularly in Section 3.3. As an aid to the use of this book, a summary of the design procedure is presented in the following table, Table 1.1

In order to enable the engineer to follow the necessary design stages, information is given in Chapter 2 on the geotechnical stability of river and channel banks and on various types of hydraulic loading. Chapter 4 gives detailed information on biotechnical and structural types of revetment, which should be complemented in the case of proprietary systems by consulting the data sheets presented in Appendix 1. Formulae for design of the various revetment types in normal current flow are also given in Chapter 4: for other flow conditions (such as wave attack, high turbulence and tidal flows) design equations and guidance are given in Chapter 2. The design of filter layers is presented in Chapter 5, and Chapters 6 and 7 deal with construction and maintenance issues, respectively. Worked examples of river and channel bed and bank protection are provided in Chapter 8 to assist the designer in the use of the recommended procedures and formulae.

In general terms, all revetments, irrespective of their particular load conditions, should be designed to withstand wind, wave and current flow attack. Therefore, Sections 2.3.1 and 2.3.2 and Chapter 4 will always be necessary for the design procedure. However, in situations of high turbulence, the formulae to be used for current attack are not those in Chapter 4, which are valid only for normal turbulence, but those given in Section 2.4.1.

Table 1.1. Summary of design procedure

Design	Chapters/Sections to be consulted	Outcome
Conceptual design	Chapter 3 — Design procedure Sub-section 3.3.1 — Conceptual design Section 2.1 — Geotechnical stability Section 2.2 — Functions of revetments Subsection 2.3.2 — Waves and rapid water level changes Sub-section 2.4.1 — High turbulence Section 3.2 — Ranges of applicability of revetments Chapter 5 — Use of granular filters and geotextiles Chapter 6 — Construction issues	Identification of most suitable kind of revetment (or strategy)
Outline design	Chapter 3 — Design procedure Subsection 3.3.2 — Outline design Sub-section 2.3.1 — Current attack Sub-section 2.3.2 — Waves and rapid water level changes Section 2.4 — Specific problems Chapter 4 — Types of revetment and design formulae Chapter 5 — Use of granular filters and geotextiles Chapter 6 — Construction issues Appendix 1	Choice of revetment(s)
Detailed engineering	Chapter 3 — Design procedure Subsection 3.3.3 — Detailed engineering Chapter 4 — Types of revetment and design formulae Chapter 2 — Stability of channel beds and banks Chapter 5 — Use of granular filters and geotextiles Chapter 6 — Construction issues Chapter 7 — Maintenance procedures Appendices 1 to 6	Detailed design of the revetment(s) and filter layer(s).

The book is completed by a series of Appendices: Appendix 1 including lists of revetment systems, both proprietary and non-proprietary, and data sheets with details of the former; Appendices 2 to 5 giving advice on specification of riprap, gabion mattresses, concrete block mattresses and geotextiles; and finally Appendix 6, presenting illustrative examples of detailing at the edges and transitions of revetment works.

Stability of channel bed and banks

2. Stability of channel bed and banks

2.1. GEOTECHNICAL STABILITY

A brief introduction is given here of the major geotechnical parameters affecting bank stability and the modes of bank failure most likely to be encountered in river engineering. This section is not intended to provide the designer with a comprehensive knowledge of soil mechanics, but to remind him or her of the geotechnical factors that are relevant to the hydraulic design of river and channel revetments. The following textbooks may prove useful to complement the information given in this section: Terzaghi and Peck (1948), Capper and Cassie (1969), Craig (1987), Berry and Reid (1987), and Hemphill and Bramley (1989, Chapter 2).

2.1.1. *Soil characteristics*

Soil is defined as a natural aggregate of mineral particles which can be separated by gentle mechanical means, as opposed to rock where the minerals are connected by strong, permanent forces. Rock banks, unless they have been badly weathered, do not require protection against flow-induced erosion and therefore only soils are considered in this section.

There are two basic types of inorganic soil:

- cohesive soils, originating from chemically unstable rocks that undergo changes at the mineral level and result in an aggregate of very fine plate-shaped particles
- granular soils, originating from the physical breakdown of relatively stable rocks and formed by more regularly shaped particles.

The differences in the ways in which these soils were formed are reflected in their properties: the behaviour of granular soils is determined by their mass energy, whereas in cohesive soils it is the surface energy of the particles that is responsible for their characteristic cohesiveness, plasticity and volumetric changes (Moffat, 1990). Organic soils, such as peat and sands comprised of shell particles, have specific properties that are not covered here.

Among the many possible soil classification systems, classification by particle size is normally very useful since it is a simple way of identifying soils for preliminary assessments and gives an indication of their likely properties. The following table, Table 2.1, presents different soil categories, drainage characteristics and the nominal particle sizes that are used to establish the limits between the categories. These nominal particle sizes are determined by sieving analysis for

Table 2.1. Soil classification according to size

Soil	Size: mm	Drainage characteristics
Clay	< 0·002 or 2μ (microns)	Impervious (intact clays) Very poor (weathered clays)
Silt	0·002–0·06	Poor
Sand	0·06–2·0	Fair
Gravel	2·0–60	Good
Cobbles	60–600	Good
Boulders	> 600	Good

granular soils (D_{50} is usually taken as the nominal size; it corresponds to the size below which 50% of particles by weight are smaller) and by sedimentation techniques for cohesive soils. It should be noted that, as this classification is based on size, it is not always absolutely logical: for example not all clay-size soils are formed by clay minerals and vice versa.

The results of sieving and sedimentation analysis are represented graphically in grading curves such as those shown in Figure 2.1. The x-axis represents the particle size in logarithmic scale and the y-axis is a natural scale giving the percentage by weight finer than the corresponding particle size. Grading curves provide information on the type of soil and on the range of particles of which the soil is composed. Soils formed by a wide range of particle sizes have gently sloping grading curves. These well-graded soils tend to have greater strength and stability than uniform or poorly graded soils, which have steeper grading curves; however, uniformly graded materials usually have good drainage characteristics.

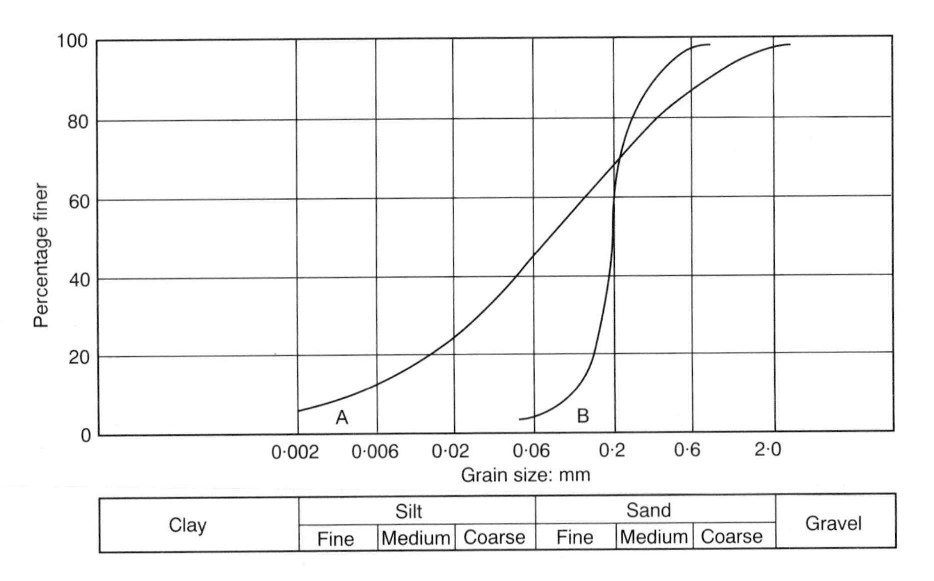

Figure 2.1. Examples of grading curves: A — well graded soil; B — uniformly graded soil

2.1.2. *Geotechnical parameters*

Soils are two- or three-phase systems consisting of solid particles, water and/or gas. Water can exist in a variety of forms in the soil structure besides the liquid phase: it can be part of the gaseous phase in the form of water vapour, or of the solid phase as absorbed, interlayer or structural water (these last two forms are only present within clay soils).

The shear strength of a soil is defined as its maximum resistance to shearing stresses, failure occurring when this is exceeded, usually taking the form of surfaces of slip. Pore water pressure has a significant influence on the soil strength since the replacement of voids by water reduces the contact between soil particles. An increase in pore water pressure will decrease the soil's shear strength, whereas a reduction will lead to enhanced strength. Rather than thinking in terms of total shear strength, it is usual to consider the effective shear strength. By subtracting the effect of the pore water pressure, this quantity gives a better measure of the soil strength. This is so because the application of stress usually results in a temporary increase in the pore pressure. The effective shear strength is given by the sum of two terms, the effective cohesion c' and the effective internal friction $\sigma'_{\mathrm{n}} \tan \phi'$:

$$\tau = c' + \sigma'_{\mathrm{n}} \tan \phi' \tag{2.1}$$

where
$c' = c - u$; c is cohesion and u is the pore water pressure
σ'_{n} is the effective normal stress
ϕ' is the effective angle of internal friction.

For cohesionless soils, such as sands and gravels, which derive their shear strength largely from internal friction, τ is given by:

$$\tau = \sigma'_{\mathrm{n}} \tan \phi' \tag{2.2}$$

since $c' = 0$.

Cohesive soils derive their shear strength both from cohesion and friction, but when drainage has had little or no time to occur, saturated clays may appear to have cohesion only. Their shear strength can therefore be calculated as:

$$\tau = c' \tag{2.3}$$

Some soils, however, fall into an intermediate category, the cohesive-frictional soils.

Representations of the shear strength against the effective normal stress were first carried out by Coulomb in the eighteenth century; stress conditions above the line defined by Equation (2.1) will cause failure within the soil, whereas stable conditions plot below that line, which is generally known as the Mohr–Coulomb line (Figure 2.2). The parameters c' and ϕ' are usually determined by laboratory testing. For a preliminary assessment of the geotechnical stability of banks, it is possible however to take values of c' and ϕ' from Table 2.2. Laboratory tests on

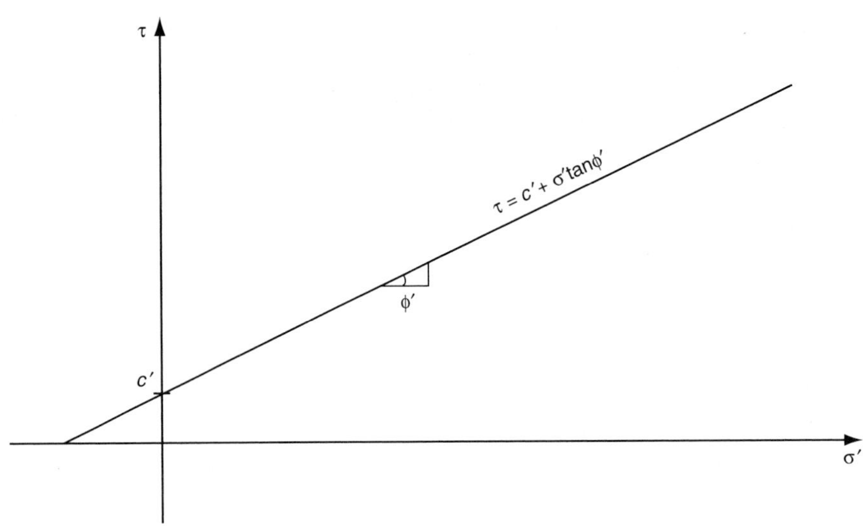

Figure 2.2. Typical Mohr–Coulomb graph (effective shear strength against effective normal stress)

undrained soil samples have shown that the values c' and ϕ' are generally very close to c and ϕ obtained from tests where the samples are sheared under conditions of full drainage.

In Table 2.2, values of the angle of internal friction are also presented for granular soils of various sizes and shapes, and for riprap. These values are approximately the same as the values of the angle of repose, which is the angle to the horizontal at which a heap of material will stand without support, for

Table 2.2. Values of cohesion and angle of internal friction

Material	Cohesion c: kN/m^2	Angle of internal friction ϕ*: °		
Clays				
Very stiff or hard	> 150			
Stiff	100–150			
Firm to stiff	75–100			
Firm	50–75			
Soft to firm	40–50			
Soft	20–40			
Very soft	< 20			
Silty sand			27–34	
Granular soils		Rounded	Rounded and angular	Angular
Particle size D_{50}				
< 1 mm		30	~33	33–35
1–10 mm		30–32	32–36	33–40
10–100 mm		32–37	33–40	~40
Riprap			40–45	

*For uncompacted sand, the angle of internal friction ϕ coincides with the angle of repose. For riprap the angle of repose is typically between 35 and 42°.

uncompacted, dry or permanently submerged granular soils. This parameter ϕ is commonly used in revetment design to account for the reduced stability of particles placed on slopes, due to the component of their weight in the direction of the slope, W_s (see Figure 2.3).

The coefficient for reduced stability (i.e. reduced critical shear stress) Ω is usually defined as:

$$\Omega = \sqrt{\left(1 - \frac{\sin^2\alpha}{\sin^2\phi}\right)} \tag{2.4}$$

where
α is the bank slope
ϕ is the angle of repose of the bank material (see Table 2.2).

The modes by which banks can collapse are many and varied (e.g. deep rotational, shallow, planar failures) and depend on a number of factors too great to describe here in detail. As mentioned above, geotechnical engineering textbooks should be consulted if the need arises; the very informative book by Hemphill and Bramley (1989, Chapter 2) is also a good source of information, as it was specifically written with the river engineer in mind. Figure 2.4 was reproduced from that publication to illustrate types of mass failure in banks.

It was noted earlier that pore water pressure is a major factor affecting soil strength. As a matter of fact, this statement can be broadened to include the weight of water itself, as water infiltrated into cracks causes not only a rise in pore water pressure but imposes an additional weight on the bank. This increases its susceptibility to collapse.

High water pressures can result from a rapid lowering of the water level in the channel that is not matched by an equally fast drainage through the bank material. This is likely to occur in banks of fine-particle soils and/or those protected by revetments of low permeability, such as for example pitched stone and solid concrete slabs. Excessive pore pressure not only reduces the shear strength of the soil, but the movement of water from the bank (seepage) can cause piping, which may ultimately lead to failure. Piping is a process by which the fine particles of the soil are washed away from the bank by seepage, leaving only the coarser particles in the soil and a trail of large voids. The soil structure

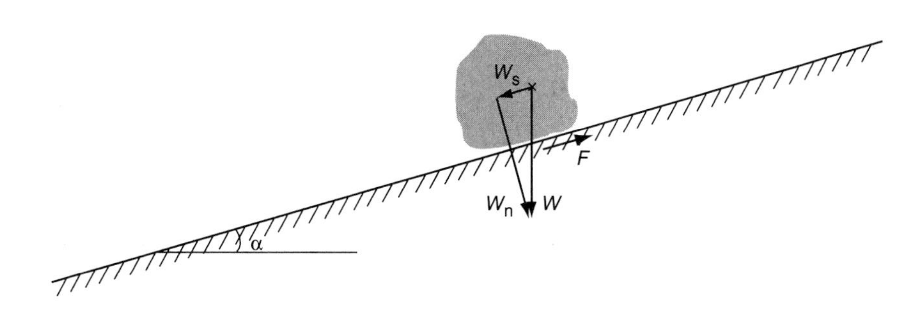

Figure 2.3. Forces acting on a particle placed on a slope

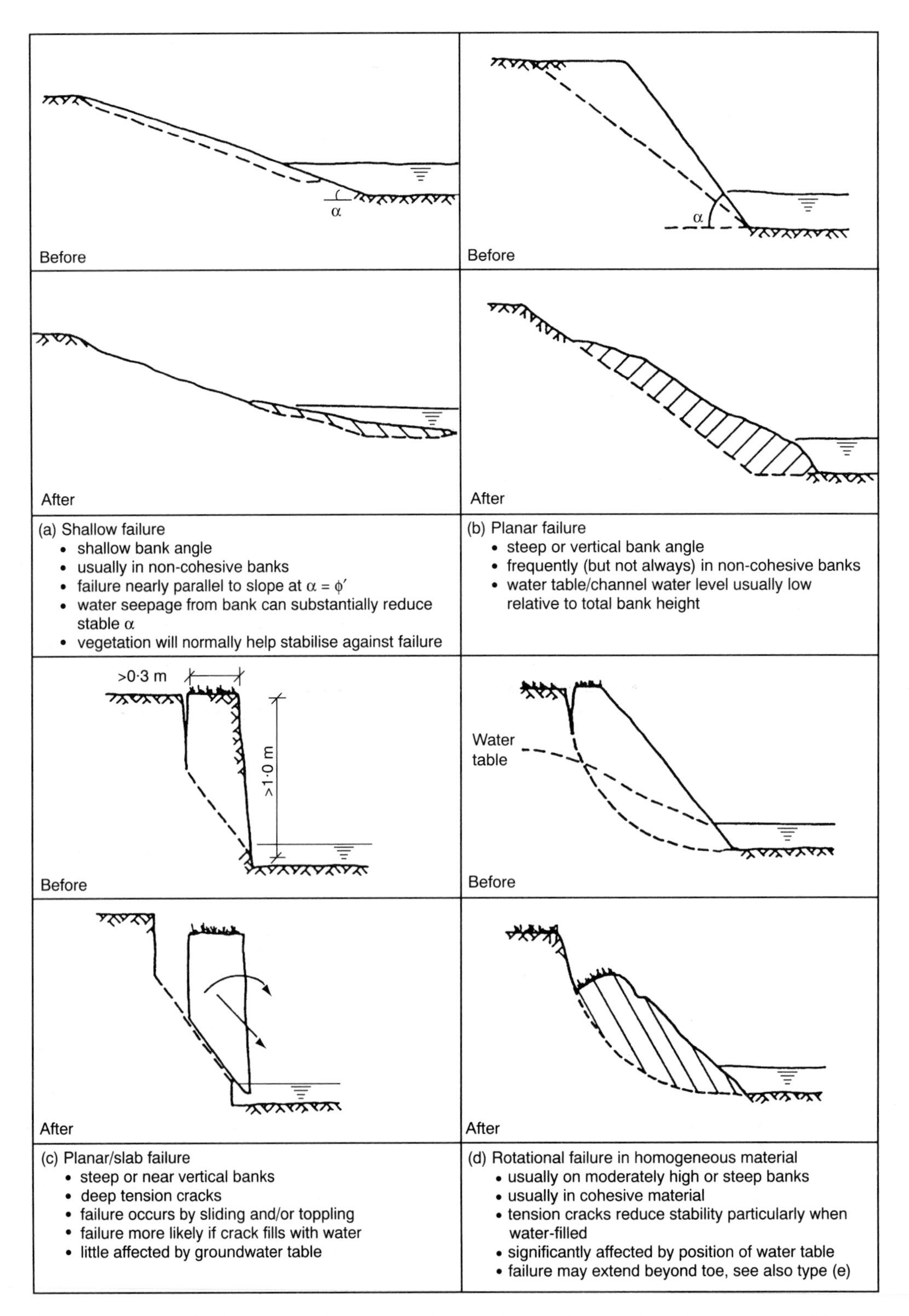

Figure 2.4. Typical types of river bank failure (from Hemphill and Bramley, 1989)

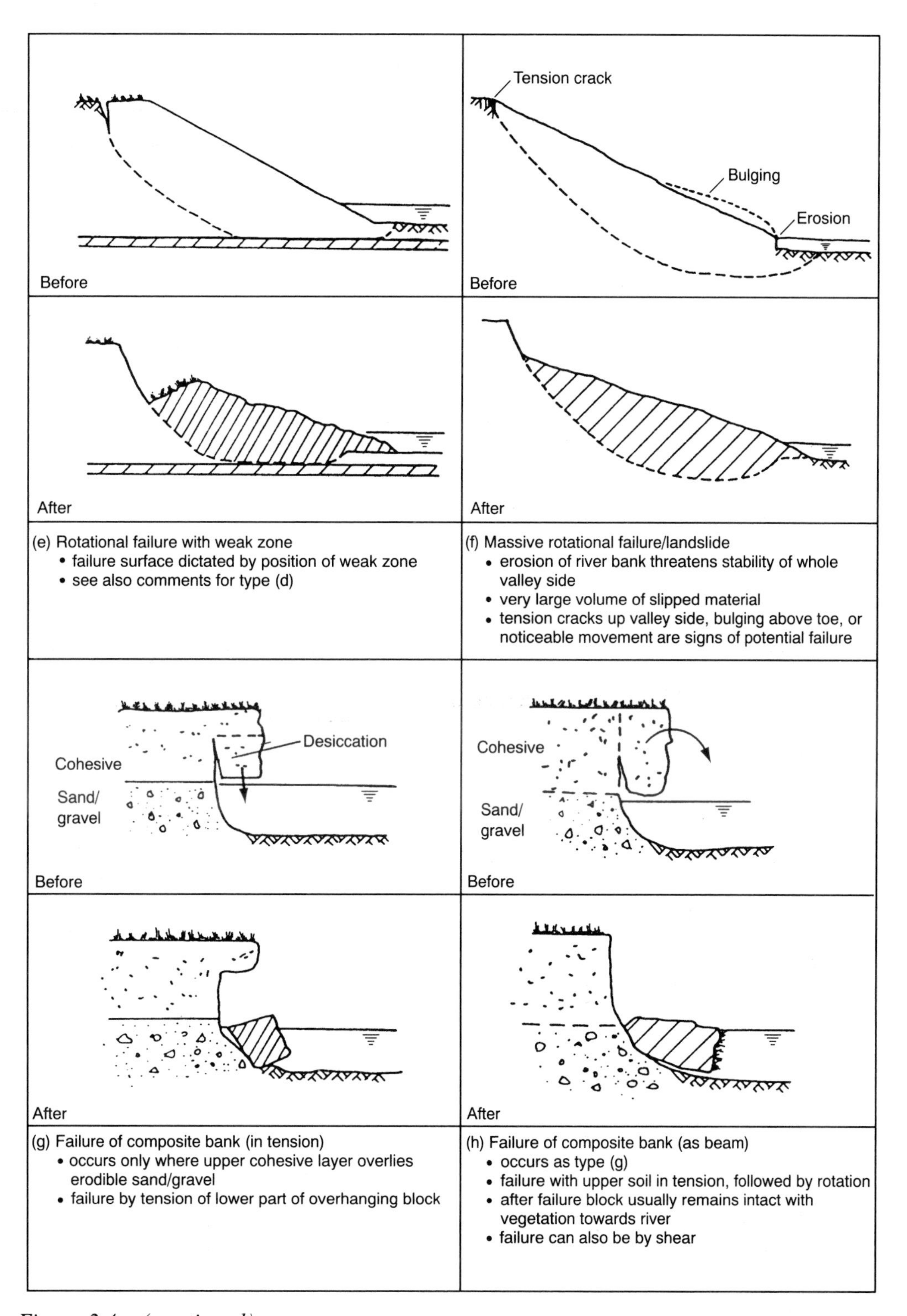

Figure 2.4 (continued)

becomes more open and also more prone to surface erosion. Soils where this process is more likely to occur are non-cohesive fine soils such as sands and silts; coarser sands will tend to drain more effectively and the colloidal bonds of cohesive soils allow them to withstand higher hydraulic gradients than granular soils.

Besides water and the nature and geometry of the bank, other factors can be responsible for bank failure:

- surcharge loading on top of the banks or on river berms (e.g. inadequate design of access roads or footpaths)
- the development of cracks due to tension or dry conditions on the top of cohesive banks, sometimes aggravated by traffic of people or animals
- the growth of large bushes or trees, which may in some cases help to stabilise banks, may also cause instability in steep banks where little soil is available for an adequate establishment of the root system.

2.2. FUNCTIONS OF REVETMENTS

From a geomorphological point of view, rivers are dynamic bodies of water not just because they convey flows from their source to their mouth, but also because their boundaries are by no means fixed. Even in rivers that have reached equilibrium (or regime rivers), instability of bed and banks is observed near the apex of meanders in the form of sediment deposition and erosion (see Section 2.4.2). This is much more apparent in non-regime rivers, where the watercourse is attempting to find its equilibrium slope and shape, which correspond to a situation where the average transport rate of sediment equals the average rate of sediment supply. Unstable rivers show variations in layout, and heavy scour and deposition of sediments at a regional scale, rather than just locally. Human attempts to stabilise these rivers using revetments in local reaches without global consideration of the problem will often fail, and therefore are not the best solution for these areas. General information on river geomorphology is given in textbooks such as Richards (1982) and guidance on assessment of bank erosion for British rivers can be found, for example, in Thorne *et al.* (1996).

Revetments are primarily built to reduce the hydraulic load acting on the soil, with the aim of preventing or halting erosion. Erosion at a local scale can be due to a number of causes and be determined by human action, as well as by animals, plants and natural physical causes (such as the erosion that occurs at river bends or in pools and riffles).

The following are some of the most common causes of human-induced erosion: straightening of river meanders; dredging of gravel and sand from river beds; removal of vegetation from banks and river islands; regulation of rivers by building weirs and other engineering works; and navigation, which introduces additional loads on beds and banks (waves and boat currents). Leisure activities like fishing and water sports, for example, can create erosion pockets or accelerate the rate of erosion, as can animal grazing and burrowing into banks. Not all of these problems are best remedied by construction of revetments since correct management practices are often more efficient and economical. A typical example of these practices is to prevent the disappearance of natural vegetation and therefore the weakening of the soil, by creating a buffer zone near the river bank. Alternatively, access to the watercourse for grazing animals can be restricted to only a few well defined and protected points along the bank.

Revetments can be used to line an entire channel or river cross-section or be limited either to the banks or to the bed. As mentioned in Sections 1.2 and 2.1, revetments are not a solution to problems caused by geotechnical instability and are not expected to have a retaining wall function.

The adequate protection of the toe of a bank is one of the major roles of revetments, as scour depths tend to be bigger at such locations. If no action is taken to limit the development of scour, the stability of the bank can be endangered. In rivers and channels where only the banks are lined there are two major ways of ensuring toe protection using flexible revetments:

- protection of the toe at sufficient depth to account for maximum scour depth predictions
- provision of a flexible revetment that will continue to protect the toe as the scour hole develops.

These methods are illustrated in Figure 2.5. The first method can take the form of a deep toe trench filled with loose stone that is able to 'launch' itself to a limited extent if scour develops further — case (a), or consists of a revetment constructed down to below bed level — case (b). When this is too costly and/or difficult to implement due to excessive scour depths, the second method is used — case (c). It usually consists of a falling apron formed by riprap or by fascine mattresses. This type of revetment is built to sit on the bed of the watercourse after construction and is designed to accommodate general scour in the bed that may occur during flood events.

It was stated above that the main function of a revetment is the prevention of erosion of the underlying soil. Particularly in canals used for water supply or irrigation, revetments also fulfil the functions of ensuring good water quality and of reducing seepage losses both from the canal to the base soil and vice versa.

In situations where there is a need for bank protection but continuous revetments are not economical and the use of groynes (or spur dikes) is not feasible due to width restrictions, another solution is sometimes adopted. It consists of series of 'hard points', which can be described as 'mini groynes' or local protrusions of revetments into the river (see Figure 2.6). These are designed to prevent erosion where they are located and to limit the erosion between hard points to an acceptable degree (CUR, 1995).

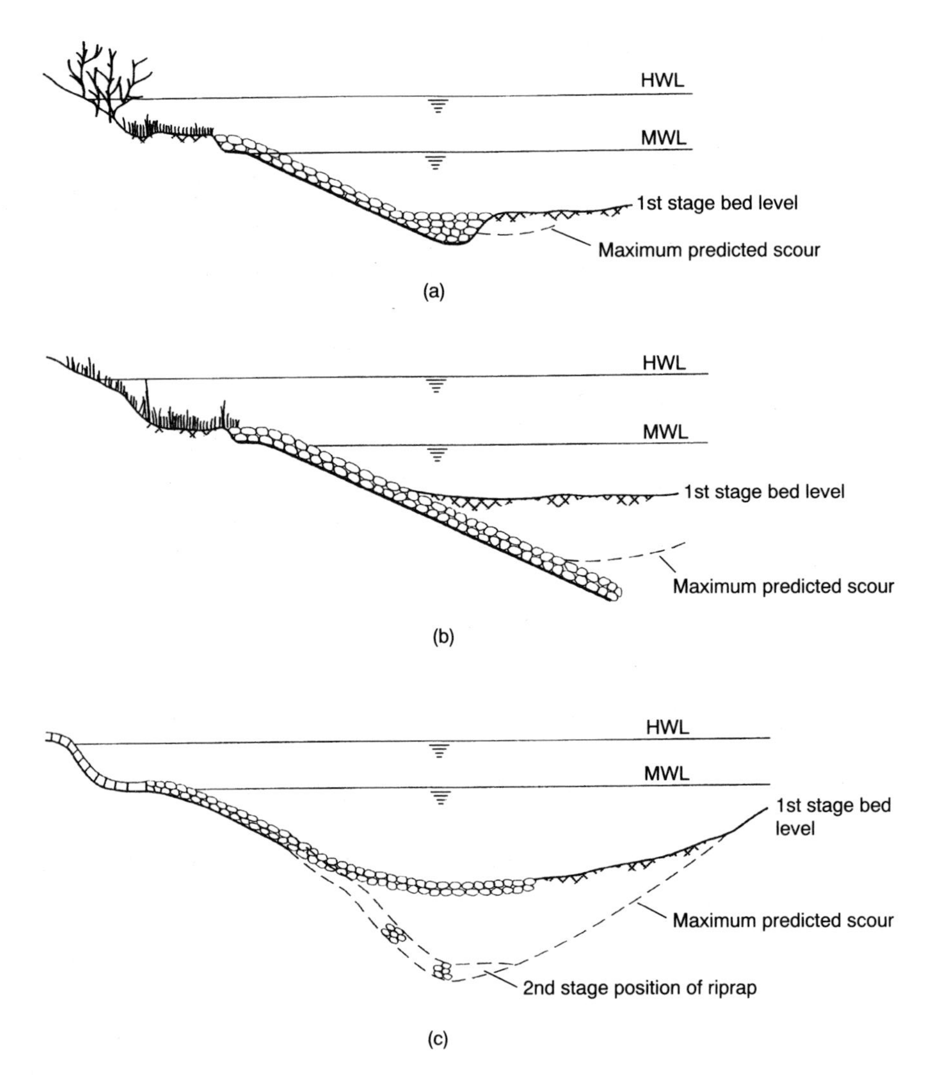

Figure 2.5. Typical examples of toe protection using riprap: (a) toe trench; (b) below bed level mattress; (c) falling apron

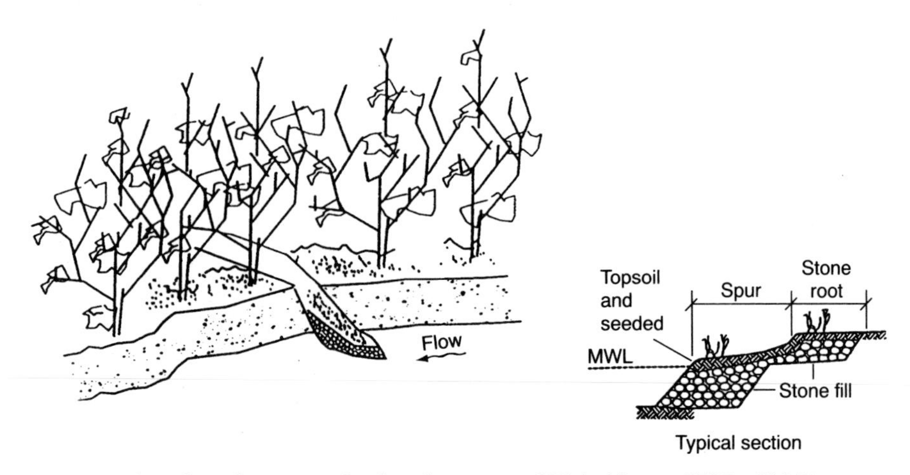

Figure 2.6. Example of application of a hard point in USA (from CUR, 1995)

2.3. HYDRAULIC LOADINGS

2.3.1. *Current attack*

Currents are the first type of hydraulic loading that needs to be taken into account when designing revetments for rivers and channels. In the context of this section, 'current attack' relates principally to fluvial, or unidirectional, flow in the streamwise direction, since this is in most situations the major destabilising force acting on a revetment. Other types of current, like those generated by boat motion are addressed later in this section, and the effect of tidal currents is dealt with in Section 2.3.3.

As well as streamwise current attack, all watercourses will also experience secondary currents. These can be a result of the cross-sectional shape of the channel, of irregularities in the channel layout (see the effect of meanders in Section 2.4.2), or of high turbulence (see Section 2.4.1). Figure 2.7 illustrates the case of a natural river cross-section. Note the formation of secondary currents near the left bank, due to a marked change in the cross-sectional topography and roughness. However, even in regular channels, like the trapezoidal channel depicted in Figure 2.8, secondary currents are still present. Examples of isovels (curves of equal velocity) in channels of various shapes are presented in Figure 2.9. These provide information on the areas of the channels cross-section where the highest velocities are likely to occur.

It is usual in hydraulic engineering to classify flows as subcritical, critical and supercritical. *Subcritical flows* have mean velocities smaller than $\sqrt{(gy)}$, where g is the acceleration due to gravity and y is the flow depth, and disturbances to the flow are propagated both upstream and downstream. This means that if, for example, the channel's natural roughness is increased by some form of bed or bank protection, increased water levels will also be observed at some distance upstream of the protected reach. Apart from upland rivers and certain upstream river reaches, most natural rivers have subcritical flows. During floods or around obstructions to the flow these conditions may change and supercritical flow can be established temporarily. *Supercritical flows* have mean velocities greater than $\sqrt{(gy)}$; disturbances in this case can only be propagated downstream. Although most channels are designed for subcritical conditions, there are cases where, due

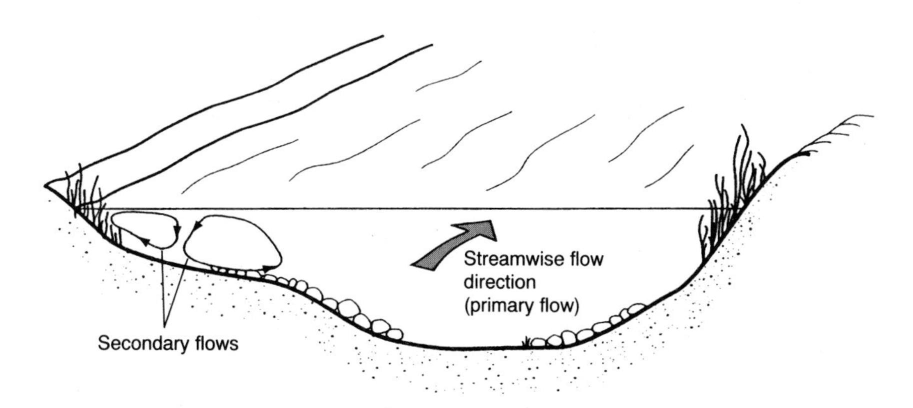

Figure 2.7. Cross-section of natural channel (adapted from Knight and Shiono, 1996)

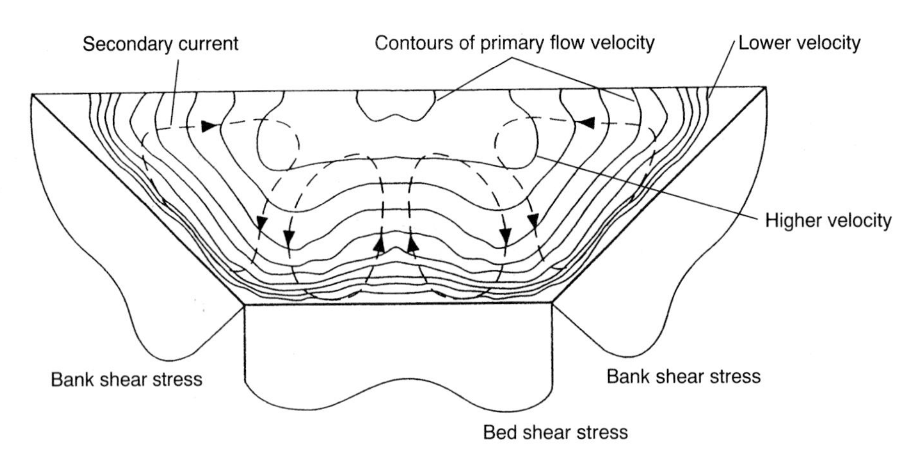

Figure 2.8. Velocity contours, secondary currents and shear stress distribution in trapezoidal channel (adapted from Knight et al., 1994)

to topographic and economic constraints, steep, supercritical channels are necessary. Examples of these include channels for the return of flow from dam spillways to the natural watercourse, temporary and permanent diversion channels, outfall channels from power stations, etc. *Critical flow* is in between these two types (mean velocity equal to $\sqrt{(gy)}$). At critical flow conditions, if the water depth is known, the discharge in the watercourse can be determined, since there is a unique dependency between them. The Froude number, a non-dimensional quantity defined as the ratio of the mean velocity to $\sqrt{(gy)}$, is a very convenient indicator of the characteristics of the flow: it is equal to 1 for critical flow, > 1 for supercritical and < 1 for subcritical flow.

As well as applying the above classification, it is also useful to define flows in terms of the levels of turbulence present. This aspect is addressed in detail in Section 2.4.1 and therefore only a basic introduction to the subject is given here. One possible way is to distinguish between what can be denoted as 'normal river turbulence' and 'high turbulence'. Normal turbulence levels are those expected in

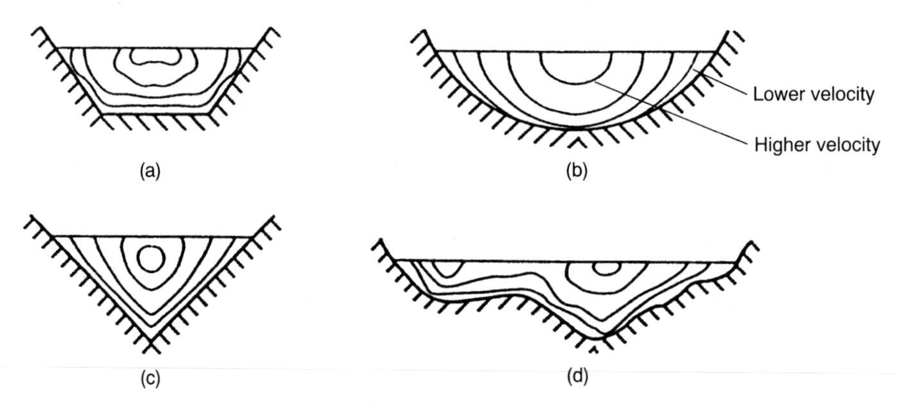

Figure 2.9. Typical curves of equal velocity (lower velocities near the channel boundary) adapted from Chow (1973): (a) trapezoidal channel; (b) shallow ditch; (c) triangular channel; (d) irregular (natural) channel

straight river reaches and in mild bends, whereas high turbulence includes a wide range of situations, from sharp meanders to the extreme case of hydraulic jumps. These are features that accompany the transition between supercritical and subcritical flows, with considerable loss of energy and turbulence generation.

Several different methods have been developed to quantify the forces imposed on river and channel boundaries by current flow and to determine the conditions for stability of such boundaries. These various methods are based, in general, on one of the following two principles:

- that stability is dependent on the shear stress at the boundaries and that a critical value should not be exceeded to avoid erosion
- that there is a critical (or maximum permissible) velocity above which stability is jeopardised.

In the shear stress approach (also known as the tractive force principle) the shear stress is defined as the average drag force per unit area of the wetted perimeter of the channel:

$$\tau_0 = \rho g R S \tag{2.5}$$

where

ρ is the water density
g is the acceleration due to gravity
R is the hydraulic radius (defined as the ratio of the wetted area and the wetted perimeter) and
S is the energy slope (equal to the channel invert slope in uniform flow).

The distribution of shear stress in channels, except in very wide rectangular ones, is not uniform, as can be seen in Figure 2.8. In trapezoidal cross-sections, shear stress is maximum near the toes of banks, which are vulnerable areas in terms of stability. If the unit tractive force imposed by the flow on the wetted boundary is below the critical value for the material forming the channel, erosion will not occur. This principle can be applied both to natural and revetted channels. Some formulae have been developed that use the shear stress principle but avoid the determination of the energy slope S, which is a difficult parameter to calculate or measure accurately. Instead of S, resistance equations such as Manning's equation, or equations for the vertical velocity profiles are used (Maynord, 1993).

The limiting velocity approach is adopted in this book because of its simplicity when compared to the shear stress approach. The mean cross-sectional velocity U is a quantity that is easy to calculate by dividing the flow by the cross-sectional area. However, research has found that it is not suitable for inclusion in stability formulae since the stability of revetment units is determined by local velocities, near the bed and banks. Although the velocity near the boundary is generally accepted as being the limiting velocity, it is sometimes difficult or impractical to measure and hard to predict as it is strongly affected by local bed features. One way of overcoming this difficulty is to use the depth-averaged velocity U_d, which can be determined from measurements of point velocities along a vertical. The formulae presented in Chapter 4 for design of revetments under current attack are given in terms of U_d. Figure 2.10 illustrates the difference between the mean cross-sectional velocity and U_d.

The following equations are given to provide the engineer with means of relating, in an approximate way, values of flow velocity for situations where field measurements are not possible and results of physical models are not available:

- Relationship between mean cross-sectional velocity U and bed (or bottom) velocity U_b—valid for rough turbulent flow, as expected in most river situations

$$U_b = \frac{U}{0{\cdot}68 \log_{10}(y/k_s) + 0{\cdot}71} \tag{2.6}$$

  where
  y is water depth
  k_s is a roughness height that can be taken as equal to D_{50} of the boundary material

- Relationship between the depth-averaged velocity U_d and the bed (or bottom) velocity U_b at 10% of the water depth above the bed. This relationship was obtained from field measurements in a UK river and is valid for normal turbulence flows in straight river reaches

$$U_b = 0{\cdot}74 \text{ to } 0{\cdot}90\ U_d \tag{2.7}$$

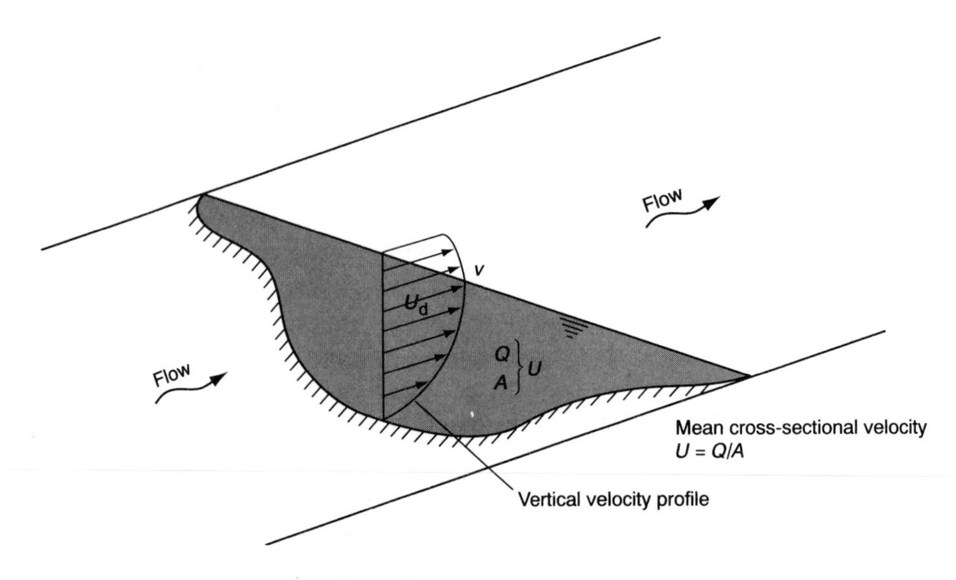

Figure 2.10. Flow velocity in channel cross-section

Ship-induced currents

Currents in navigable rivers and canals may also be a result of ship movement. The water motion produced by boats is complex, involving changes in water level, waves and currents of various degrees of magnitude, and can have a severe effect on the stability of beds and banks. Waves are, in the great majority of cases, the most critical of these loadings and are described in some detail in the next section, Section 2.3.2, where the general patterns of boat-induced water motion are also presented. The present section gives a brief description of boat-generated currents.

Two different types of current are produced by ship movement:

- Return currents — these are parallel to the channel banks (assuming that the boat is not sailing eccentrically) but of opposite direction to the vessel motion
- Propulsion-induced currents (or screw-race) — these are high velocity jets generated by the ship's propeller and can have various orientations from parallel to normal in relation to the waterway boundaries.

Return currents are only generated during the time in which the vessel is passing but can cause high shear stress on the bed and banks of the waterway. They are basically a function of the vessel's speed, the water levels produced by the vessel's motion and the relative wetted cross-sectional areas of the ship and the channel. Design methods for the determination of return current speeds can be found in PIANC (1987). Based on existing research, Hemphill and Bramley (1989) have suggested the indicative values for UK rivers given in Table 2.3.

Propulsion-induced currents only have a significant effect on the stability of the waterway boundaries when they arise from starting vessel movement from a stationary position or during manoeuvring operations. Damage to boundaries therefore tends to occur in front of locks and mooring piers and near banks. The degree of erosion is greater the heavier the propulsion system and the longer the time the jets impinge on the boundaries. Therefore, the lower the ship's speed, the greater the potential for erosion (and the consequent formation of bars). Depending on the position of the propeller relative to the boundaries, the erosion can be due to shear stress (when the axis of the propeller is parallel to the surface of the boundary), solely due to hydrodynamic pressures (when the axis is normal to the boundary as near vertical banks) or to a combination of both.

In PIANC (1987) an equation is suggested for the calculation of the bottom velocity U_b due to propeller jets for vessels starting from rest (SI units should be used):

Table 2.3. Return currents due to ship motion

Type of watercourse	Return current velocity: m/s
Smaller canals	< 1
Larger canals	$< 1{\cdot}5$
Navigable rivers	2-3

$$U_b = \alpha \times 1{\cdot}15\left(\frac{P_d}{D_o^2}\right)^{0{\cdot}33}\frac{D_o}{Z_b} \qquad (2.8)$$

where

α takes a value between 0·25 and 0·75, depending on ship and rudder types
P_d is the engine power in kW
D_o is given by

$D_o = D_p$ for ships with propeller in a nozzle
$D_o = 0{\cdot}7D_p$ for ships without nozzle
where D_p is the propeller diameter

Z_b is the vertical height from the boundary to the propeller axis.

In the absence of specific data to introduce in the above equation, U_b can be taken as 2·5 m/s as a first approximation.

For design of revetments in navigable waterways it is recommended that the current velocities generated by ship movement be checked and compared with other hydraulic loadings. However, it is likely that this comparison will show that boat-wave attack is more detrimental to the stability of revetments. Since ship-induced loadings are very much dependent on the types of vessel that cruise the waterways, field measurements are the best means of assessing the flow conditions that they produce.

Wind-induced currents

Currents can also be generated by wind blowing over water surfaces, which can be significant for large expanses of water and if the wind speed is sustained for a certain amount of time. In steady state conditions, the current velocity can be estimated to be about 2–3% of the wind speed (Hedges, 1990). However, wind currents are unlikely to reach these values for the situations considered in this book and therefore they can usually be neglected in the design of river revetment systems.

2.3.2. Waves and rapid water level changes

As well as fluvial current attack, the bed and banks of rivers and channels may also be subjected to forces due to waves. These can be generated by different processes: by wind blowing over long stretches of water (wind-induced waves), by the movement of boats and ships (boat wash), by sudden releases of water from power stations and reservoirs, by flash floods or surges. The effect of these types of wave is most noticeable at the banks rather than at the bed, since some are essentially surface features that impinge on the sides of the channels. The stability of natural or inadequately protected banks is often jeopardised by waves, particularly those produced by boats, as shown in Figure 2.11. Rapidly varying water levels can be particularly severe in certain types of climate (e.g. those

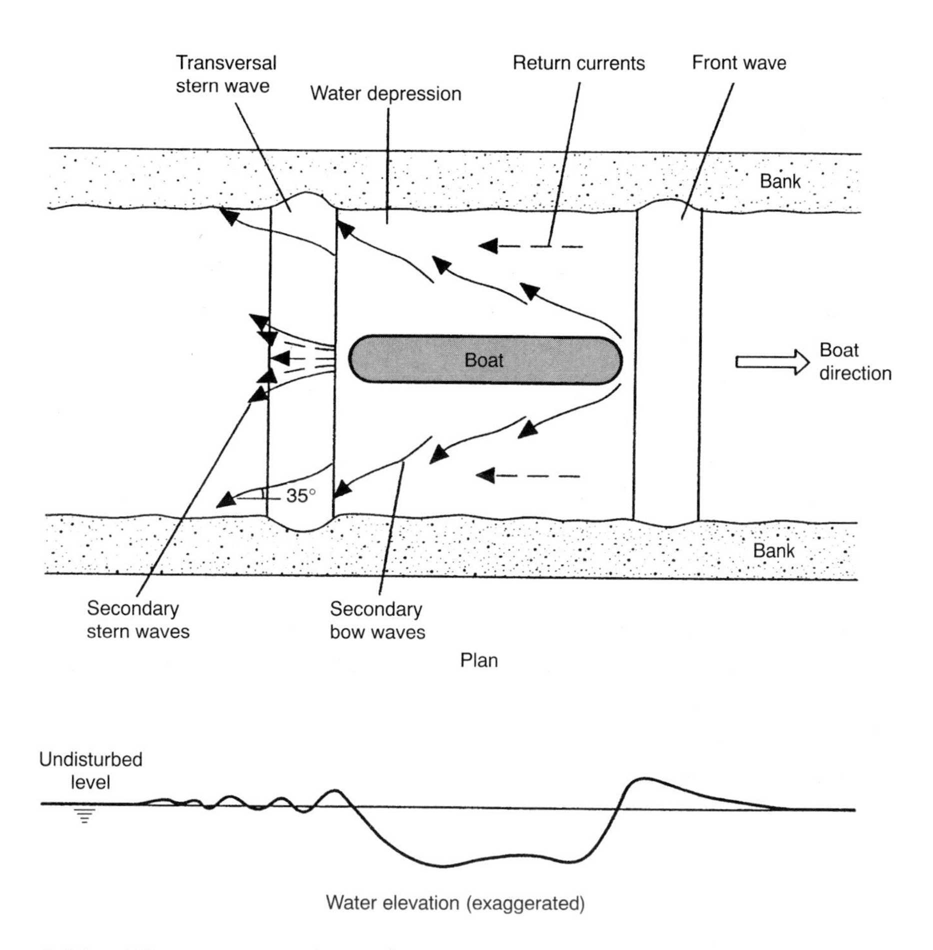

Figure 2.11. Water patterns due to boat movement

producing wadi flows) and in areas with pumped drainage schemes. As well as producing wave fronts that can destabilise the banks directly, they are usually also associated with rapid drawdown. Seepage problems due to drawdown are addressed later in Section 2.3.3.

Two main techniques can be used to determine the wave environment at a site:

- spectral analysis (in which the water surface is considered to be a random collection of waves of different heights and periods) or
- the determination of the significant wave height H_s with period T_z (in which the whole spectrum is deterministically equivalent to regular waves of the same energy).

Statistically, the significant wave height represents the average wave height (crest to trough) of the highest third of waves, in a given period. The period T_z is the mean zero-crossing time found from the analysis of wave records. This second approach is commonly adopted in river engineering because of its simplicity.

Wind-induced waves

Wind waves in inland waterways are random, surface gravity waves whose significant height and period depend essentially on the wind speed and direction and on the size of watercourse over which the wind is blowing (the fetch, F). Saville *et al.* (1962) presented a method for determining the effective fetch length to use for the calculation of the wave characteristics. Some typical specific situations that result from application of Saville's method are summarised in Figure 2.12. In the context of this book, which excludes large water bodies such as reservoirs and estuaries, fetches much above, say, 500 m are unlikely to occur. This, and the maximum wind speeds expected on site, will limit the amplitude of wind-generated waves that can be expected and generally cause them to be less severe than boat wash, for example. For a more accurate estimation of the fetch in UK bodies of water, see Yarde *et al.* (1996). The procedure recommended for design of revetments subjected to wind waves is given in Figure 2.13 and is described below.

The significant wave height H_s and period T_z can be calculated using the simplified Sverdrup-Munk-Bretschneider (SMB) equations (in Hemphill and Bramley, 1989):

$$H_s = 0{\cdot}00354(U_{10}^2/g)^{0{\cdot}58}F^{0{\cdot}42} \qquad (2.9)$$

$$T_z = 0{\cdot}581(FU_{10}^2/g^3)^{0{\cdot}25} \qquad (2.10)$$

where

U_{10} is the wind speed at a height of 10 m above mean water level (obtained from local records or Met. Office data)
F is the fetch (see Figure 2.12)
g is the acceleration due to gravity.

Data from UK inland sites indicate that the maximum values of U_{10} are typically between 18 m/s (sheltered sites) and 25 m/s (exposed sites). Although the significant wave height is representative of the wave environment at the site, the stability of revetments is dependent on the maximum wave height H_i. The following relationship can be used to calculate H_i:

$$H_i = 1{\cdot}3H_s \qquad (2.11)$$

The formulae given next are recommended for design of riprap and concrete blocks in moderate wave conditions (see Hemphill and Bramley, 1989). In very harsh wave environments (i.e. breaking and plunging waves or exceptionally high waves) it is suggested that reference be made to more comprehensive publications such as van der Meer (1988) or PIANC (1987). Independent information on other types of revetment is given in PIANC (1987) but some useful and specific design guidelines may also be available from manufacturers' literature.

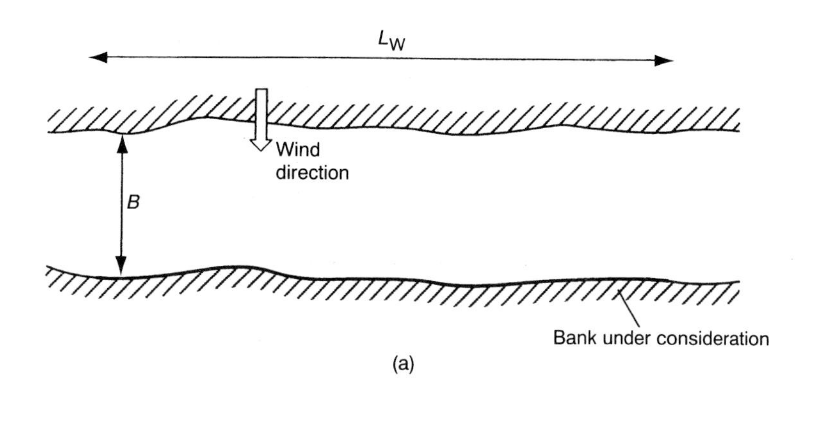

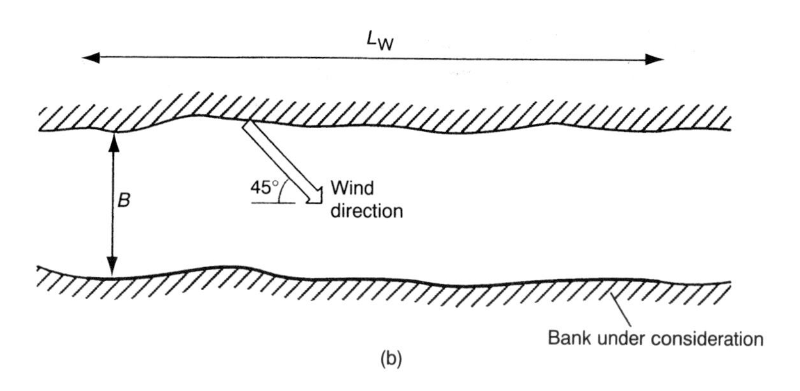

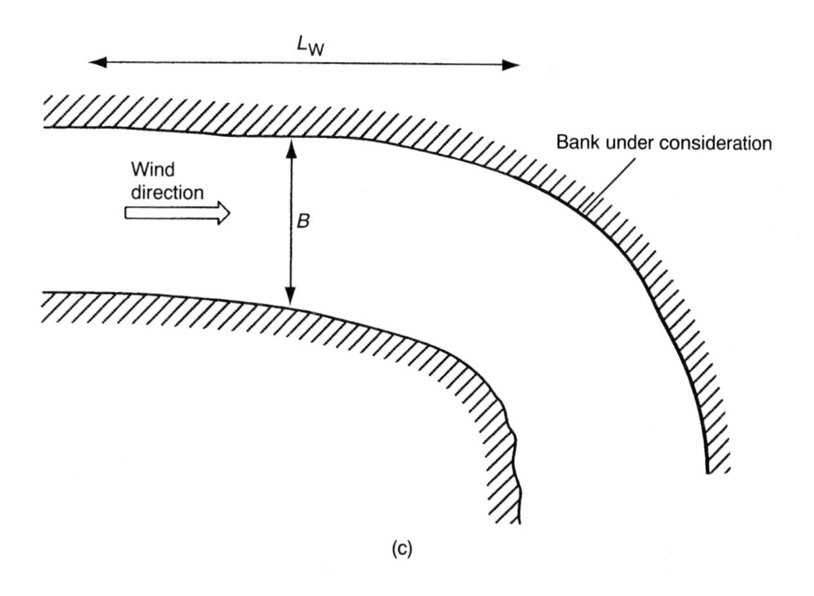

Figure 2.12. Determination of fetch according to Saville et al. (1962): (a) $F = B$; (b) $F = 2{\cdot}5B$ (for $L_W > 20B$); (c) $F = (3L_W + 67B)/40$

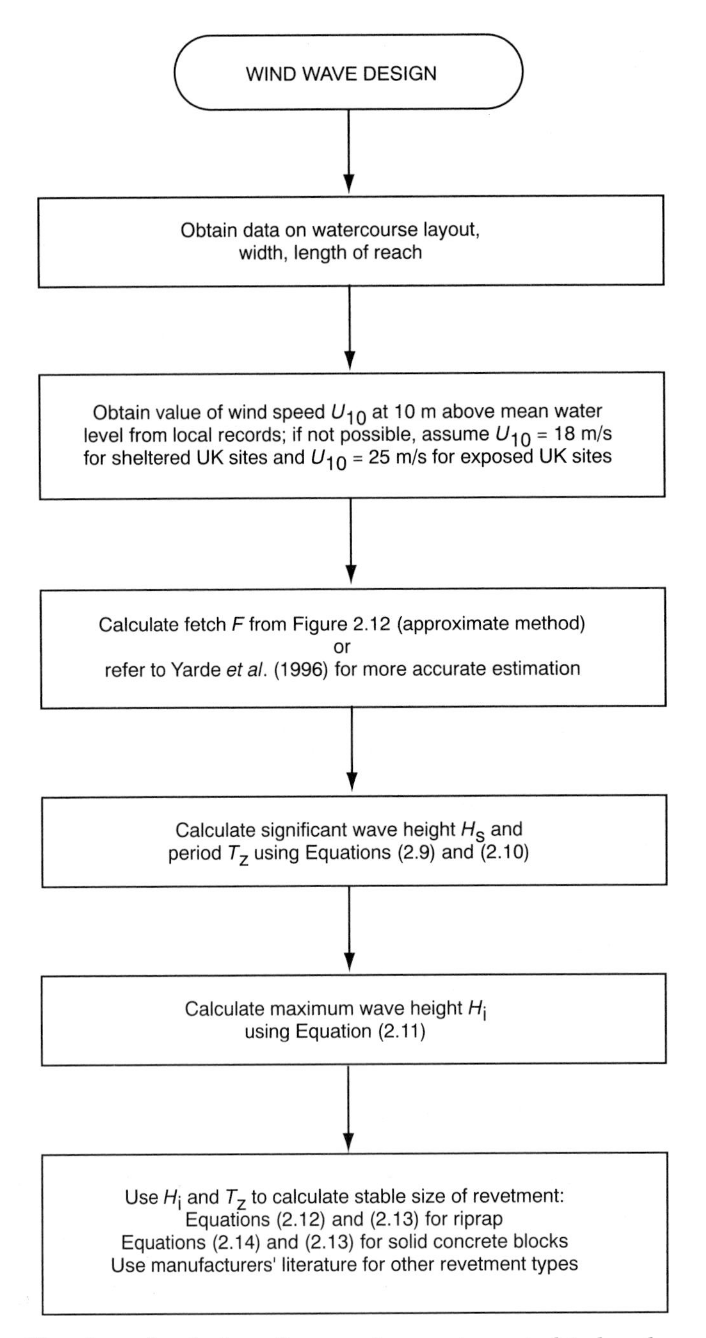

Figure 2.13. Flowchart for design of protection against wind-induced waves

Riprap design (see also Section 4.1.1)

$$D_{n50} = 0{\cdot}34 \frac{H_i}{s - 1} I_r^{0{\cdot}5} \tag{2.12}$$

where

D_{n50} is the stone size
H_i is the maximum wave height
s is the relative density of the material, defined as ρ_s/ρ, where ρ_s is the density of the stone and ρ is the density of water
I_r is the Iribarren number, which can be taken in the following simplified form

$$I_r = \frac{\tan \alpha}{\left[(2\pi H_i)/(1{\cdot}3 g T_z^2)\right]^{0{\cdot}5}} \tag{2.13}$$

where α is the bank slope and g is acceleration due to gravity. Physically, the Iribarren number represents the ratio of the slope of the bank to the steepness of the incident waves.

Concrete block design—plain, solid blocks

$$D = G \frac{H_i}{(s - 1)\cos \alpha} I_r^{0{\cdot}5} \tag{2.14}$$

where

D is the block thickness
H_i, s, α and I_r are defined as above
G is a coefficient dependent on the type of blocks

$G = 0{\cdot}19$ to $0{\cdot}26$ for loose blocks
$G = 0{\cdot}15$ to $0{\cdot}19$ for cabled block mats.

The above equation does not cover the whole range of block systems available, many of which are cellular and therefore have quite different wave resistance characteristics. Specific design data should be sought for other types of block, preferably from laboratory tests or from their manufacturers.

Boat wash

The movement of boats in rivers and channels produces a complex pattern of waves and currents, with associated changes in water level, as illustrated in Figure 2.11 and summarised in Table 2.4.

As a vessel moves along a watercourse, it displaces water from the bow. In water bodies that are constrained within fixed boundaries, two main water patterns have been identified: a primary system, which is associated with the general motion of water around the boat, and a secondary system, which is

Table 2.4. Flow patterns due to boat movement

Types	Water pattern	Description
Primary pattern (due to the general flow around the boat)	Front wave	Wave transverse to the channel, just ahead of the boat
	Water depression	Reduction in water level between the front and stern waves caused by return currents
	Return currents	Induced currents in opposite direction to the boat movement
	Transversal stern wave	Wave transverse to the channel, just behind the boat
Secondary pattern (surface disturbances)	Bow waves	Oblique waves propagating from the bow of the boat
	Stern waves	Oblique waves propagating from the stern of the boat
Screw-race (propeller-induced)	Propeller jets	Currents produced by ship propellers

formed by oblique waves generated at the bow and stern of the vessel. At a particular moment in time the primary system is formed by a surface elevation ahead of the boat (front wave) which then falls sharply to a water level depression; this is followed by a rapid rise in level at the stern (transversal stern wave). Return currents with direction opposite to that of the boat motion are responsible for the depression between bow and stern waves. The secondary system is formed by transverse and diverging waves with interference peaks that have a distinctive V-shaped pattern, which depends essentially on the boat speed and the channel water depth. For most situations the angle of the individual wave crests to the line of the bank can be taken as 35°.

Boat wash is the term generally used to describe the water movements generated by boat motion. Some information was given in Section 2.3.1 concerning return and propeller-induced currents, and therefore the present section will deal only with boat wash. In navigable UK watercourses, this is probably the most severe kind of loading that is responsible for erosion of banks. The rapid drawdown following the passage of waves may generate excessive pore pressures, and high speed boats may produce breaking stern waves and additional flow turbulence.

The flow patterns generated by boat wash depend on a number of factors. These are principally related to the boat geometry (in particular the boat length and also its shape and width), the boat speed, the waterway geometry and the water depth in the channel. An important parameter is the ratio of the cross-sectional boat area below the waterline and the undisturbed cross-sectional area of the flow in the channel. This ratio, the relative blockage factor, is associated with the amplitude of the primary boat waves and the limiting boat speeds that need to be imposed in order to minimise erosion potential.

Monitoring of wave amplitudes and periods generated by the types of vessel that use a particular waterway is undoubtedly the best approach on which to base the design of bank protection. When this is not feasible, the values of wave

Table 2.5. Typical values of wave height for UK waterways

Channel type	Boat size: t	Wave height: m
Smaller canals	< 80	< 0·3
Larger canals	< 400	< 0·5
Navigable rivers	< 40	< 0·4

height given in Hemphill and Bramley (1989) reproduced in Table 2.5 can be used as a first approximation for UK waterways.

Hemphill and Bramley (1989) also presents the following general design equations for stability of riprap in boat-generated waves.

Riprap design (see also Section 4.1.1)

- Primary waves (transverse stern waves)

$$D_{n50} = 0{\cdot}67 \frac{H_i}{(s-1)(\cot\alpha)^{1/3}} \tag{2.15}$$

- Secondary waves

$$D_{n50} = 0{\cdot}56 \frac{H_i}{(s-1)}(\cos\beta)^{0{\cdot}5} \tag{2.16}$$

where
D_{n50} is the stone size
H_i is the height of the highest waves expected (in the absence of specific data, use the values suggested in Table 2.5)
s is the relative density of stone, defined as ρ_s/ρ, where ρ_s is the density of the stone and ρ is the density of water
α is the angle of bank slope to the horizontal
β is the angle of the individual wave crests to the line of the bank; usually taken as 35°.

Stability formulae for other types of revetment may be obtained from research studies described in manufacturers' literature or from laboratory tests. The particular case of concrete block stability has been well investigated and found to depend on the block's weight, size, shape and interlocking or cabling method, as well as on the permeabilities of the block revetment and of the sublayer. PIANC (1987) suggests a method that takes into account these factors, which is recommended for design. However, the following equation, which was derived from comparison of stability between riprap and concrete blocks, may be used as a first approximation to give the stable size of solid plain blocks against the attack of secondary boat waves: $D = 0{\cdot}43\, H_i (\cos\alpha)(\cos\beta)^{0{\cdot}5}/(s-1)$.

Wave run-up

It should also be appreciated that in the presence of waves, revetments need to be designed so that bank protection is ensured up to the maximum level reached by wave run-up. This level, higher than the still water level, depends on the slope of the bank and the roughness of the revetment. In general terms, rough revetments such as riprap, gabion mattresses and some types of concrete block mats may reduce wave run-up by up to about 40% when compared with smooth revetments such as concrete or asphalt linings. Berms also have a reduction effect: above a certain optimum width, the wider the berm the more efficient it will be. Formulae for calculation of wave run-up can be found in PIANC (1987). Rules of thumb for first estimation give the height above still water level as $2H$ for riprap and $4H$ for concrete mattresses or asphalt mats, where H is the design wave height (see also Appendix 6).

2.3.3. Tidal flows and other differential loads

Tidal flows

The downstream reaches of watercourses that flow towards the sea will experience the effect of tides to a greater or lesser extent. In enclosed seas like the Mediterranean this effect is very small and can usually be neglected when designing river bed and bank revetments. However, in other regions of the world, tides can be responsible for large variations in water level (in places as high as ± 6 m around mean sea level), and for currents that alternate in direction about every 6 hours. The length of natural watercourses under tidal influence is dependent essentially on their configuration and on the amplitude of the tides; in regulated rivers this is also true to some extent, and an attempt is normally made to maintain some of the natural flow conditions. Tidal reaches of several tens of

Figure 2.14. Example of tidal river experiencing bank erosion

kilometres are common in the UK and elsewhere. Figure 2.14 illustrates the case of tidal rivers suffering bank erosion (note the silt covered banks, characteristic of this type of river reach).

The effect of tides should be considered when designing revetments because of the following main reasons.

- The variation in water levels caused by tides implies two things: that *protection of upper banks is generally necessary*, and that *excessive hydrostatic pressures* may build up behind revetments when the water level is falling. Unlike non-tidal reaches, where the upper bank is often designed to a lesser degree of protection because it is only sporadically flooded, in tidal reaches the same degree of protection is usually ensured up to the high water level. The second aspect concerns the build-up of pore pressures behind revetments which, although sometimes more serious for faster water level variations such as those caused by waves, also needs to be considered in the design of tidal revetments. It is important to realise that if the permeabilities of the revetment and/or filter are small when compared with the base soil, uplift forces will be generated by excess pore pressure. These forces need to be matched by the weight of the revetment and filter, otherwise collapse of the revetment will occur, as well as deformation and loss of soil from the banks and bed. To avoid this situation, a suitable design of the revetment filter layer is absolutely essential (see Chapter 5). In order to avoid leaching out of materials from finer soil layers through coarser soil layers, PIANC (1987) recommends, as a rule of thumb, that the permeability of each layer (including the cover layer) should be 20 to 50 times greater than the permeability of the underlayers (including the base soil). Whenever possible, permeable revetments should be chosen for tidal reaches; if, however, revetments of low permeability cannot be avoided, provision should be made for an adequate number of weepholes to enable the relief of hydrostatic pressures that tend to accumulate behind the revetment.
- The alternation in flow direction during tidal cycles will form flood and ebb channels in the bed of alluvial rivers. These are likely to shift with time from one location to another and to vary in shape and depth, thus creating potential difficulties and uncertainties in ensuring adequate bed protection. It is therefore important to commission detailed surveys at frequent intervals.
- Whereas in unidirectional river flow the protection of river beds and banks in the vicinity of structures is in many cases confined to the downstream reach, in tidal watercourses this protection is required both upstream and downstream. The extent of revetments is therefore usually greater, and care should be taken to design adequate edge details at both ends of the revetment.
- Tidal reaches provide habitats for flora and fauna that differ from other river environments in that the water can be saline and sediments tend to be fine grained. Silts and muds are common and, because they are transported in suspension in high tides, they are likely to deposit on the upper banks of revetments during tide recession. Molluscs and slimes are often found to establish on silt-covered riprap and other types of protection, giving a more natural appearance to the revetted area.

For the design of revetments under tidal flows it is recommended to adopt the formulae given in Chapter 4, which are valid for current attack, unless waves and turbulence are significant in the tidal reaches under consideration. In this latter case, refer to Sections 2.3.2 and 2.4.1. For both cases, the above guidelines should be followed to help to achieve a suitable design.

It is worth noting that peak tidal velocities tend to occur around mean tide level, but consideration of high tide levels is obviously required for the design of the upper river banks. The engineer should also consider the probability of high tide occurring with flood flow to obtain the maximum velocity.

Other differential loads

The discharge of excess water from reservoirs and sudden increases in river flow during flood conditions are other examples that may cause erosion of river beds and banks. These are essentially random processes that are difficult to predict or estimate in general terms; they will need to be addressed on an individual basis. In these cases the designer of river revetments should be aware not only of the destabilising action of the currents generated, but also of another potentially destabilising action: a difference in the magnitude of loads acting behind and in front of the revetment. Usually known as differential loads, they encompass a wide range of situations, some of them addressed in Section 2.1 where information was given on geotechnical parameters affecting revetment design.

Differential loads can be generated when there is a substantial difference between the permeability of the base soil and that of the revetment. A typical example occurs with the use of impermeable revetments on coarse base soils without provision for adequate release of the pore pressure that can build up behind the revetment. Even in uniform flow conditions and in the absence of waves, the differential pressure and resulting uplift forces may cause failure of the revetment. Differential loads may also be purely external, as is the case of varying water levels, particularly when the levels drop in short periods of time and the base soil and revetment have to respond accordingly. Steep hydraulic gradients across the thickness of the revetment are generated, which may lead to geotechnical instability (see Section 2.1).

2.4. SPECIFIC PROBLEMS

2.4.1. High turbulence

In hydraulics, flows are classified into two categories, laminar and turbulent, but most geophysical flows like those occurring in rivers and channels are turbulent. However, the distinction can be made between the levels of turbulence that occur in fairly straight river sections and the higher levels that result predominantly from disturbances to the normal flow pattern. The first type is denoted here 'normal flow' and the second 'high turbulence flow'.

Turbulence is a process by which the energy of an 'orderly' steady flow is converted into the random kinetic energy of eddies of decreasing sizes down to the molecular level (Yuen and Fraser, 1979). The very fast and irregular movement of the fluid particles produces instantaneous changes both in the intensity and direction of the flow velocity. Information on the mechanism of turbulence and relevant formulae can be found in a wide range of publications, with variable degrees of complexity (see, for example, Escarameia and May, 1992, Tennekes and Lumley, 1972, Nezu and Nakagawa, 1993).

The importance of considering turbulence in river engineering resides in its effect on water levels, sediment transport, water surface disturbance and the forces it imposes on the boundaries. For the design of river revetments it is essential to take account of these forces as they can be responsible for erosion of natural banks and beds and for the destabilisation of revetments.

Turbulence generation (or an increase in its intensity) is a means by which the flow readjusts itself back to a regular pattern after being subjected to a local feature impinging on the flow. Highly turbulent flows can be found in a number of situations, as listed below and illustrated in Figure 2.15:

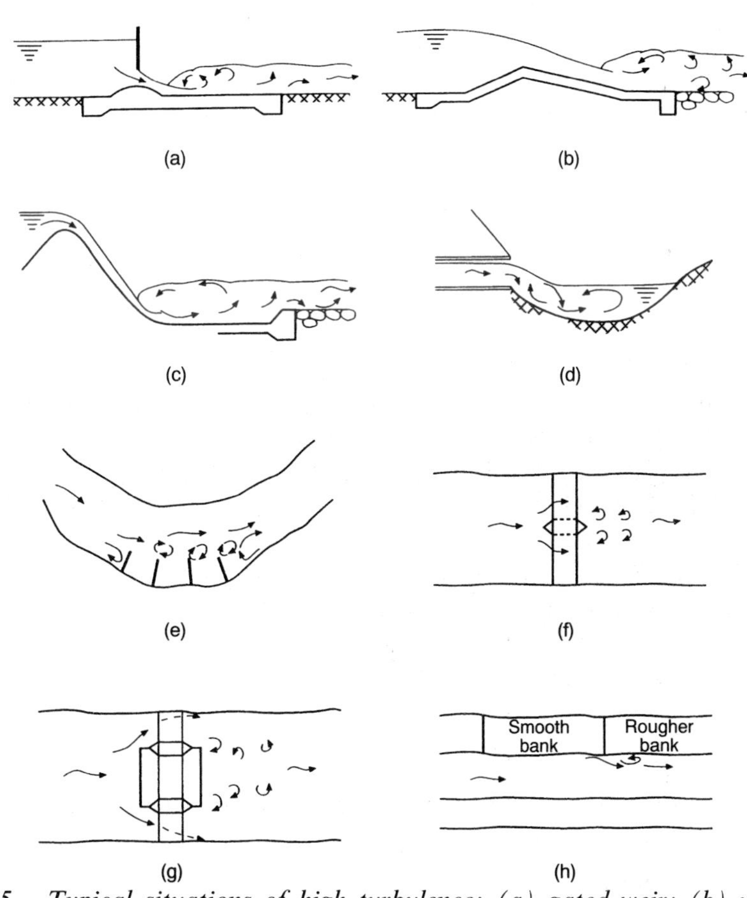

Figure 2.15. Typical situations of high turbulence: (a) gated weir; (b) ungated weir; (c) spillway and stilling basin; (d) culvert (discharging perpendicularly to a stream); (e) groynes (plan view); (f) bridge piers (plan view); (g) cofferdam (plan view); (h) transitions (plan view)

- downstream of hydraulic structures
  - gated and ungated weirs and barrages (see Figure 2.16)
  - spillways and stilling basins
  - culverts
- near river training structures such as groynes
- around bridge piers, cofferdams and caissons
- at the transition from a very smooth lined bank to a natural (rougher) bank or much rougher type of revetment
- at sharp bends and other abrupt changes in flow direction
- at variations in bed level.

Some of the situations depicted in Figure 2.15 show the formation of a hydraulic jump in the case of the weirs and the spillway, although jumps can also occur in some of the other cases. This is a stationary wave that establishes the transition between supercritical (or rapid) flow and subcritical (or tranquil) flow as described in Section 2.3.1. A hydraulic jump is the most dramatic form of high turbulence, particularly if the jump is not submerged. The dissipation of energy that occurs in a jump can actually be observed at the water surface, which looks very irregular with rolling, and sometimes bursting, eddies. This activity at the surface is an indication of what the bed is experiencing: random fluctuations of pressure and velocity, that are occasionally extremely violent.

Excessively high turbulence levels can persist even at considerable distances downstream of the front of the jump, and therefore they should be taken into account when designing bed and bank protection. Downstream of obstructions to the flow such as bridge piers or groynes, the increase in turbulence is not as easy to detect as in a hydraulic jump; sometimes, however, vortices and other flow irregularities can be seen at the water surface.

Figure 2.16. Example of high turbulence downstream of a small weir

The level of turbulence present in the flow can be measured quantitatively in more than one way, but it is common to define it in terms of turbulence intensity. This is basically a ratio that reflects the variation of flow velocity around the mean in relation to either the mean flow velocity or the shear velocity. Quantification of the turbulence intensity requires data that are not always available. Therefore, in many cases turbulence has been defined only in a qualitative way. It is common to find turbulence given in subjective terms such as high, medium or low (or normal), without any indication of how to assess it. Field measurements have recently been carried out in the River Thames to determine what levels of turbulence are present in various types of river situation in the UK (Escarameia *et al.*, 1995). The turbulence intensity TI was defined quantitatively as the ratio of the 'root mean square' of the streamwise velocity component and the time-averaged velocity also in the streamwise direction. Both quantities refer to 10% of the water depth above the bed. In the absence of site-specific information, it is recommended that the values in Table 2.6 be considered for design purposes.

Among the large number of stability equations that have been developed over the years (almost all for the design of riprap) there are three that should be considered because of the allowance they make for the turbulence of the flow. The equation due to Escarameia and May (1992) is a form of the well known Izbash equation and was developed from laboratory tests carried out at HR Wallingford, UK, on riprap, concrete blocks and gabion mattresses. This is the only equation known where turbulence is included in a quantifiable way. Work developed in the Netherlands on stability of riprap and mattresses (both stone and concrete block mattresses) led to the Pilarczyk (1990) equation. It includes several empirical coefficients derived in part from prototype trials, among which there is a turbulence factor which varies in accordance with the type of flow

Table 2.6. Turbulence levels

Situation	Turbulence level	
	Qualitative	Turbulence intensity TI
Straight river or channel reaches and wide natural bends ($R/W > 26$)*	Normal (low)	0·12
Edges of revetments in straight reaches	Normal (higher)	0·20
Bridge piers, caissons and groynes; transitions	Medium to high	0·35–0·50[†]
Downstream of hydraulic structures (weirs, culverts, stilling basins)	Very high	0·60[‡]

* R—centreline radius of bend; W—water surface width at the upstream end of the bend (see Section 2.4.2).

[†] The lower limit should be used when protecting across the width of the river or channel whereas the upper limit refers to local protection around piers or groynes.

[‡] *Important note:* this value refers to turbulence levels persisting downstream of hydraulic structures or of stilling basins and concrete aprons, where these are present; the value therefore does not apply to sections very close to large weirs or spillways not provided with energy dissipation structures.

condition. The third equation is part of the US Army Corps of Engineers' Design Procedure and was developed and refined by Maynord over a number of years. This equation was specifically developed for riprap following extensive laboratory studies carried out in small- and full-scale conditions. It should be noted, however, that the failure criterion behind this equation is different from that of the other equations. Instead of using the threshold of movement, the criterion adopted by Maynord was that the underlying material should not be exposed to the hydraulic load. This was intended to account for the thickness of the riprap and for segregation during placing, but it means that the design formula will incorporate a smaller margin of safety against failure than the previous ones. For this reason, it is recommended here that a higher safety factor than the minimum proposed by Maynord be adopted.

The three equations considered above are presented next. Most of these equations are given in terms of depth-averaged velocity U_d—see Section 2.3.1 for conversion to other velocity values. More information on these equations, their development and applications is given in Escarameia and May (1992), Escarameia (1995), CUR (1995), Thorne *et al.* (1995), Maynord (1993), US Army Corps of Engineers (1981).

Design equations

1. Escarameia and May (1992)—HR Wallingford

Recommended for design of *riprap*, *loose* or *interlocking concrete blocks* and *gabion mattresses*

$$D_{n50} = C\frac{U_b^2}{2g(s-1)} \tag{2.17}$$

where

D_{n50} is the characteristic size of stone
D_{n50} is the size of the equivalent cube ($= (W_{50}/\rho_s)^{1/3}$)

W_{50} is weight of particle, ρ_s is density of stone

C is a coefficient that takes account of the turbulence intensity TI.

Values of TI for various flows conditions are given in Table 2.6. TI is defined as (rms $u)/\bar{u}$, where rms stands for 'root mean square', u is the streamline velocity component and $\bar{u}$ is the time-averaged velocity also in the streamwise direction, both measured at a point 10% of the flow depth above the bed. Values for C are shown in Table 2.7.

g is the acceleration due to gravity
s is the relative density of the revetment material, defined as ρ_s/ρ.
U_b is the velocity near the bed (at 10% of the water depth above the bed).

For the design of bank protection, U_b should refer to the toe of the bank.

Table 2.7. Values of C for use in Escarameia and May's equation

Type of revetment	Value of C	Observations
Riprap	$12 \cdot 3TI - 0 \cdot 20$	Valid for $TI \geq 0 \cdot 05$ and for design of bed and bank protection on slopes of $1V{:}2H$ or flatter
Loose (or interlocking) concrete blocks	$9 \cdot 22TI - 0 \cdot 15$	Valid for $TI \geq 0 \cdot 05$ and for design of bed and bank protection on slopes of $1V{:}2 \cdot 5H$ or flatter
Gabion mattresses	$12 \cdot 3\ TI - 1 \cdot 65$	Valid for $TI \geq 0 \cdot 12$ and for design of bed and bank protection on slopes of $1V{:}2H$ or flatter

For turbulence intensities $TI \leq 0 \cdot 50$ a relationship between $U_{\rm b}$ and the depth-averaged velocity $U_{\rm d}$ was obtained from field tests and can be used if values of $U_{\rm b}$ are not available:

$$U_{\rm b} = (-1 \cdot 48TI + 1 \cdot 04)U_{\rm d} \qquad \text{for } TI \leq 0 \cdot 50 \tag{2.18}$$

For turbulence intensities $TI > 0 \cdot 50$ the relationship between $U_{\rm b}$ and $U_{\rm d}$ is greatly affected by random shedding of eddies and therefore only a provisional equation can be mentioned: $U_b = (-1 \cdot 48TI + 1 \cdot 36)U_{\rm d}$.

Important note: Equation (2.17) includes a safety factor which was determined so that the equation gave conservative estimates of stable size for all the experimental data on which it was based.

2. Pilarczyk (1990)

Recommended for design of *riprap, cabled concrete blocks, box gabions* and *asphalt mattresses.*

$$D = \frac{\phi}{\Delta} \frac{0 \cdot 035}{\psi_{\rm cr}} K_{\rm T} K_{\rm h} K_{\rm s}^{-1} \frac{U_d^2}{2g} \tag{2.19}$$

where

D is the characteristic size of the protection (see Table 2.8)

ϕ is a stability correction factor (see Table 2.8)

Δ is the relative density of the revetment (see Table 2.8)

$\psi_{\rm cr}$ is a stability factor (see Table 2.8)

$K_{\rm T}$ is a turbulence factor (note that it is different from the turbulence intensity TI)

$K_{\rm T} = 1 \cdot 0$ for normal river turbulence

$K_{\rm T} = 1 \cdot 5$ to $2 \cdot 0$ for high turbulence (e.g. downstream of stilling basins, local disturbances, sharp outer bends)

$K_{\rm h}$ is the depth factor

For high turbulence flows it is likely that the velocity profile will not be fully developed and therefore it is suggested that the following equation be used:

$$K_{\rm h} = \left(\frac{D}{y}\right)^{0 \cdot 2} \tag{2.20}$$

where y is the water depth. For bank protection, take y at the toe of the bank.

Table 2.8. Values of coefficients in Pilarczyk's equation

Parameter in Equation (19)	Riprap	Box gabions and gabion mattressess	Cabled blocks and asphalt mats	Observations
D	D_{n50}	D_{n50}	Thickness	
ϕ	0·75 1·0–1·5	0·75 1·0–1·5	0·50 1·0	Continuous protection At edges and transitions
Δ	$s-1$	$(1-n)\ (s-1)$	$(1-n)(s-1)$	s is relative density n is porosity ($n \approx 0{\cdot}4$ for stone and sand)
ψ_{cr}	0·035	0·07	0·07	–

K_s is the slope factor defined as the product of a side slope term (k_d) and a longitudinal slope (k_l), i.e. $K_s = k_d\, k_l$; see CUR Report 169 (1995):

$$k_d = \cos\alpha\sqrt{\left[1-\left(\frac{\tan\alpha}{\tan\phi}\right)^2\right]} \quad \text{and } k_l = \frac{\sin(\phi-\beta)}{\sin\phi} \tag{2.21}$$

where α is the bank slope, ϕ is the internal friction angle of the particles (see Section 2.1.2, Table 2.2) and β is the angle to the horizontal of the longitudinal slope of the channel.

U_d is the depth-averaged flow velocity.

Equation (2.19) is an iterative equation and therefore requires a first estimate of the characteristic size of protection. The iterative procedure should continue until the assumed value introduced in the equation produces a similar value for D.

3. Maynord (1993)—US Army Corps of Engineers

Recommended for design of *riprap*.

$$D_{30} = S_f C_s C_v C_T y\left[\left(\frac{1}{s-1}\right)^{0{\cdot}5}\frac{U_d}{\sqrt{(K_1 g y)}}\right]^{2{\cdot}5} \tag{2.22}$$

where

D_{30} is the characteristic riprap size of which 30% is finer by weight—see Section 4.1.1 for relationships between D_{30} and D_{50} and D_n.

S_f is a safety factor (minimum value recommended by Maynord is 1·1, but a higher safety factor of 1·5 is suggested here for design).

C_s is a stability coefficient

C_s = 0·3 for angular rock
C_s = 0·375 for rounded rock.

C_v is a velocity distribution coefficient (which can reflect the flow turbulence)

C_v = 1·0 for straight channels
C_v = 1·25 downstream of concrete structures and at end of dykes (or groynes).

C_T is a blanket thickness coefficient (it is recommended here that C_T = 1·0 be taken for standard design; see Maynord (1993) otherwise).
y is the local water depth—for the design of bank protection take y as the water depth at the toe of the bank.
s is the relative density of stone, defined as ρ_s/ρ.
U_d is the depth-averaged flow velocity—for the design of bank protection take water depth at the toe of the bank.
K_l is a side slope correction factor

$$K_l = -0{\cdot}672 + 1{\cdot}492\cot(\alpha) - 0{\cdot}449\cot^2(\alpha) + 0{\cdot}045\cot^3(\alpha) \qquad (2.23)$$

where α is the angle of the bank to the horizontal.

g is the acceleration due to gravity.

2.4.2. *Flow around bends*

Natural rivers have a tendency to meander, i.e. to develop successive bends or meanders, particularly in the gentler slopes at the approach to the river mouth. The complexity of this geo-morphological process is not entirely understood but it seems plausible that obstructions to the flow, such as hard ground, cause the river to divert from a straight alignment. Also, if the bed slope is too steep for an equilibrium regime (i.e. where the average transport rate of sediment equals the average rate of sediment supply), the river meanders to increase its length and thereby reduce its gradient.

Once the inception of bend-forming has occurred, the meandering process will tend to sustain itself. The resulting curved flow will generate secondary flow patterns, which will cause scouring on the outside of a bend. This action, combined with sediment deposition on the inside of the bend, will help to increase the curvature of the meander. The disturbance caused by bends can persist for considerable distances downstream and is a combination of secondary currents and, in some cases, the onset of waves. The waves can cause erosion by overtopping of the banks downstream of bends, particularly in supercritical flows. For more detailed information on the hydraulics of flow around bends see, for example, Henderson (1966).

Amongst the types of obstruction or disturbance that can induce the formation of bends are:

- the build-up of a central sediment bar (which can first cause the division of the flow into two branches and later the supremacy of one in relation to the other)
- small scour holes in one of the river banks
- fallen trees on the river bed.

Erosion of the outside of bends is, in many instances, a consequence of human intervention, such as the construction of inadequately designed river training works or insufficient bank protection around hydraulic structures. These cases are likely to give rise to local excessive acceleration of the flow velocity. As in straight river reaches, damage caused to banks due to fishing purposes and overgrazing can also speed up the erosion process by weakening banks that are already exposed to non-parallel flow and secondary currents. Erosion on either side of the bank has been found to be triggered by the above factors (NRA, undated).

In the past, artificial channels and channelised rivers in urban areas were designed with the prime objective of maximising the conveyance, i.e. the flow-carrying capacity of the channel. Today, equal emphasis is placed on ensuring that artificial channels will convey the necessary flow while having a high environmental value. One way to achieve this is to incorporate features such as bends to enhance the natural appearance of the channel and promote flora and fauna habitats.

Bends can therefore be considered as desirable features that are present in both natural and built channels but it is important to realise that the flow patterns in bends can differ considerably from those found in straight reaches. Figure 2.17 presents some important features that can be identified in meandering channels:

- the thalweg, or line of maximum depth, which is found close to the outside of bends, swinging from one side of the channel to the other
- deposition of sediment in the form of point bars, which occurs on the inside of bends

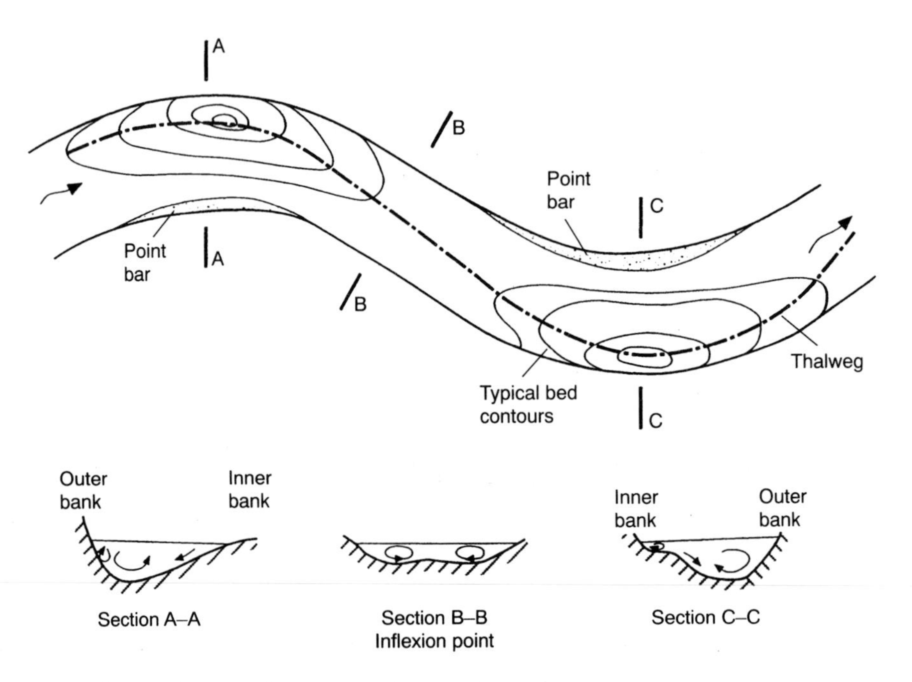

Figure 2.17. Typical example of meandering channel

- the inflexion point (Section B in Figure 2.17), which is where the thalweg crosses the centreline of the channel. It should be noted that the maximum velocity occurs approximately in the thalweg and that its position shifts with the flow discharge, sometimes coming close to the inner side of the bend. It is advisable, particularly for larger bank protection schemes, to investigate the location of the thalweg before carrying out the works.

The effect of bends on the velocity pattern is illustrated in Figure 2.18 for the case of a bend in the River Severn, UK. The velocity contours and bed profile at the inflexion point are fairly representative of sections of straight channels. In this particular meander, flow velocities are higher and closer to the outer bank at the apex section than at the inflexion point; also, the non-uniformity of the isovels (lines of equal velocity) is more apparent at the apex, which indicates the presence of stronger secondary circulation. The sharper the bend, the stronger the effect on the flow velocity patterns. This characteristic is usually defined by the ratio R/W, where R is the centreline radius of the bend and W is the water surface width. For the purpose of revetment design, only sharp bends (defined below as $R/W < 26$) need to be considered. It should be noted that this is an approximate criterion, and that physical model studies may be required for the design of complex schemes.

Not much information is available on the stability of proprietary types of revetment placed in bends. However, extensive studies have been carried out in the USA that provide guidance on design of riprap protection. The US Army Corps of Engineers' Design Procedure gives an equation (Equation 2.22) for sizing stone that includes a number of empirical coefficients to take account of several relevant parameters. Among these is the velocity distribution coefficient

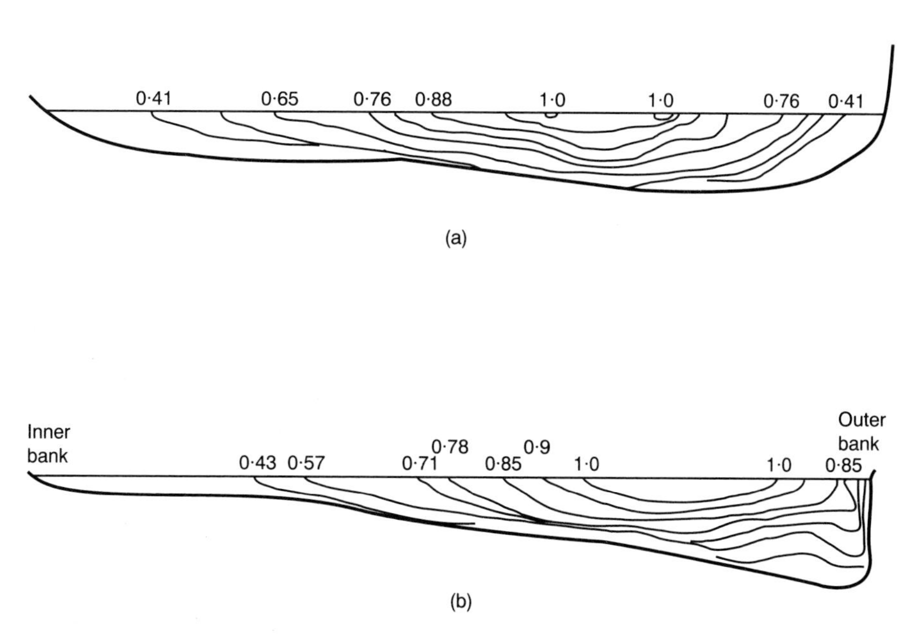

Figure 2.18. Velocity contours at the inflexion point and apex of bend in the River Severn, UK. Velocity values given as ratios of mean local to maximum velocity (adapted from Bathurst, 1979): (a) inflexion point, $Q = 17{\cdot}64\,m^3/s$; (b) bend apex, $Q = 15{\cdot}48\,m^3/s$

C_v which takes the value 1 for straight channels and for the inside of bends. In these situations laboratory studies have indicated vertical velocity profiles following a power law; the profiles became more uniform in height at the outer side of bends. Because of this increase in velocity near the bed (probably caused by secondary currents), there is greater potential for riprap movement. C_v can be calculated as follows (Maynord, 1993):

$$C_v = 1{\cdot}283 - 0{\cdot}2 \log_{10}(R/W) \tag{2.24}$$

and

$$C_v = 1 \quad \text{for } R/W \geq 26 \tag{2.25}$$

where
R is the centreline radius of the bend
W is the water surface width at the upstream end of the bend.

Depending on the geometry of the bend, this coefficient will typically increase the size of the necessary stone or blocks by up to about 30%, when compared with conditions in straight channels. Although this coefficient was derived for the US Army Corps of Engineers' Design Procedure, in view of the lack of other suitable formulae, its use is suggested for application with the other design equations presented in Chapter 4.

Care should be taken in the application of this coefficient for bends in highly turbulent environments. The coefficient was introduced to reflect non-standard velocity profiles, which can be due to a number of causes, one of them being high turbulence. Where flows are very turbulent (for example downstream of hydraulic structures) the approach recommended in Section 2.4.1 should be followed since it is likely that the destabilising effect of turbulence will override that of bends.

Based on work developed in bends with $R/W = 2{\cdot}3$, the US Army Corps of Engineers (1981) recommends that protection should be extended upstream to a minimum of one mean water surface width and downstream to a length 1·5 times the mean water surface width.

2.4.3. *Scour around structures*

Encounters between the flow and obstructions to its motion such as bridge piers, groynes (or spur dykes) and bridge abutments, result in marked changes in the vertical velocity profile and in the level of turbulence of the flow. In alluvial rivers and channels these changes, in turn, are likely to generate erosion of the following types:

- erosion due to the increase in velocity resulting from a reduction in cross-section imposed by the structure
- localised erosion that is produced directly by the presence of obstructions in the flow path.

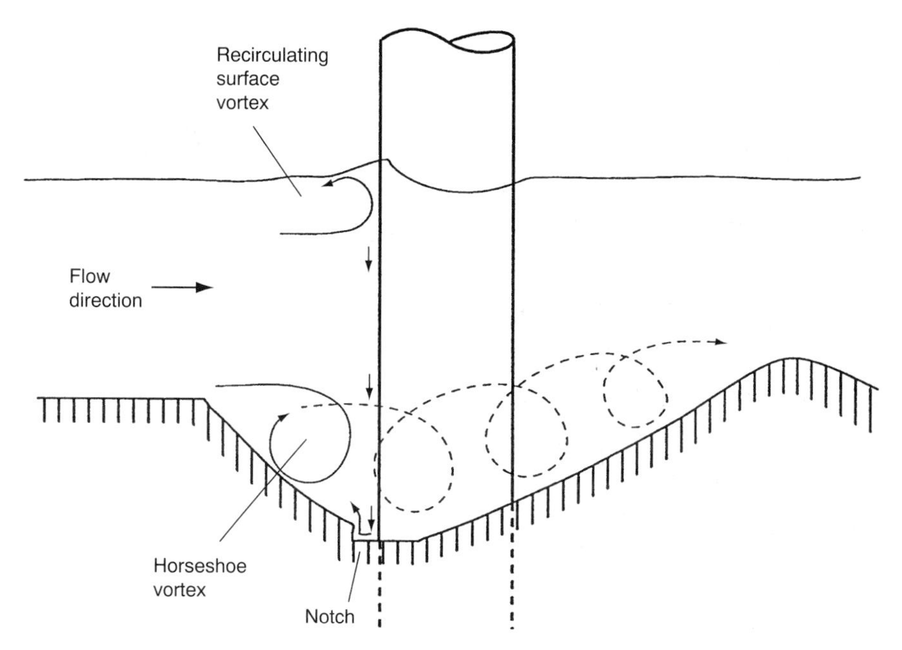

Figure 2.19. Illustration of flow patterns around structures (from May and Willoughby, 1990)

Although the resulting flow pattern is dependent on the shape of the obstruction, some flow features have been observed that are common to most cases, whether the obstruction is a circular bridge pier or an abutment, for example. These features, which are illustrated in Figure 2.19, include an upstream surface roller, descending flow along the face of the structure and wake and horseshoe vortices.

A comprehensive description of the complex processes of scour generation and development is beyond the scope of this book, but it is important to point out a few aspects. The first aspect is that erosion and ultimate scour depths around structures depend on a wide number of parameters that need to be considered in design: the flow depth and velocity, the shape, size and orientation of the structure and the bed sediment are the main ones in current flows.

In situations where waves or tides are present, the complexity of scour problems is further enhanced by the limited information available in the existing literature. This is also the case for local scouring in rivers and channels with cohesive beds, as most studies have dealt with granular materials. It should also be pointed out that, as mentioned earlier, local scour is in part produced by an increase in the turbulence of the flow. Section 2.4.1 deals with the effect that high levels of turbulence have on stability of river revetments and some limited guidance is given there for protection near bridge piers.

Protection of bridge piers

One way of protecting the foundations of bridge piers consists of preventing the development of local scour holes around piers by means of an apron. The apron may be formed by riprap, gabion mattresses, concrete block mattresses or other suitable material.

Neill (1973) recommends that the apron be laid below the expected general scour level. As a guideline for design, the extent of the apron around the pier should be approximately 1·5 times the pier width. In the case of a riprap mattress, the thickness should not be less than twice the D_{50} of the stone.

The sizing of the apron material can be carried out using Equations (2.17), (2.19) or (2.22) recommended in Section 2.4.1. When using Equation (2.17), Table 2.6 should be consulted for the choice of turbulence intensity *TI*. The value of flow velocity for use in the protection of bridge piers is the velocity through a span of the bridge. When using Equations (2.19) and (2.22), it is important to note that, in general, local scour starts at about half the threshold velocity of the sediment in the undisturbed bed upstream of the pier. For this reason, the value of flow velocity to be used in the equations should be twice the mean cross-sectional velocity upstream of the pier.

For complex situations the reader is recommended to consult other publications (for example, Neill, 1973, and Breusers and Raudkivi, 1991), which give formulae for the estimation of likely scour depths around structures and guidance on measures to limit scour development. For the particular case of cofferdams or caissons, which are large obstructions in shallow water (with ratios of water depth to structure width usually of 1 or less), see May and Willoughby (1990). In many cases, however, it is advisable to conduct physical model tests to predict scour depths with some confidence.

2.4.4. *Combined loads*

In most situations, the river engineer has to deal with a combination of different loadings, such as currents and waves, due to wind and/or boat movement. The joint action of these loads imposes destabilising forces on revetments that are likely to be more severe than any of the individual forces. However, design information on the sizing of revetments subjected to combined loads is currently not available (at least for the conditions encountered in rivers), and this is an area that requires future research.

The way in which different hydraulic loadings combine is a complex subject, more accurately dealt with by probabilistic methods of revetment design since the probability of loads occurring at the same time is an important factor. Although this book adopts a deterministic approach, some guidance is nonetheless given here for sizing of revetments under combinations of currents and waves. An approximate analysis was carried out in which the effect of waves on stability of revetments was considered to be represented by an equivalent flow velocity. This was defined as $c(gH)^{0 \cdot 5}$, where c is a numerical constant, H is the design wave height and g is the acceleration due to gravity. Consideration of the resultant of the fluid forces due to the combined waves and currents led to the following approximate guideline:

The stable size of a revetment (e.g. D_{n50} of stone) under the simultaneous action of currents and waves should be determined by adding the stable size for current attack (D_c) to the chosen stable size for wave attack (D_{bw}, for boat waves or D_{ww} for wind waves)

- *Case A*. If boat wave height is bigger than wind wave height

$$D_{combined} = D_c + D_{bw} \tag{2.26}$$

- *Case B*. If boat wave height is smaller than wind wave height

$$D_{combined} = D_c + D_{ww} \tag{2.27}$$

In Chapter 3 the flowchart Figure 3.1 and the worked examples in Chapter 8 illustrate this guideline. Whenever possible, consideration should be given to the likelihood of the simultaneous occurrence of loads.

Design procedure

3. Design procedure

The design of river and channel revetments, although dependent to a degree on the type and magnitude of the study, will normally be comprised of the following three stages:

- conceptual design
- outline design
- detailed engineering.

These stages are normally necessary before a suitable solution is reached, even if they are not formally acknowledged. In fact, smaller scheme designs or fairly routine revetment work may be carried out without specifically going through the whole procedure, but the mental process behind the three stages is generally present.

The design stages are described in detail in Sections 3.3 to 3.5 and therefore only a brief introduction is given here. The *conceptual design stage* is concerned with the evaluation of the state of the river or channel reach under consideration, the likely effect of the various possible strategies for erosion control and the selection, in broad terms, of the kind of revetment (or management practice) most suitable. This stage is followed by the *outline design stage* during which detailed design data should be collected and the predominant loads should be defined. Consideration of environmental and other factors should then lead to the comparison of alternative revetments and to the choice of solution. The process concludes with the *detailed engineering stage*, where the cover layer and underlayers (including filters) are designed and specified in detail. This stage should also include an analysis of expected costs for the life of the revetment and a detailed maintenance programme.

There are two main alternative approaches to the design of river revetment works: deterministic and probabilistic. In the first, the worst conditions of loading are determined and the revetment system is designed to withstand such loads with a certain margin of safety. This is a simpler but usually more conservative approach than the approach based on probabilistic considerations. Probabilistic design requires a statistical analysis of the various loads, i.e. the estimation of the probability of occurrence of loads and combinations of loads that may lead to the failure of the revetment. This approach involves the consideration of several scenarios; its higher complexity renders it more suitable for major protection schemes, particularly those involving extreme wave heights. In most river situations, a deterministic design is usually suitable.

The design procedure laid down in this book therefore follows a deterministic approach. The formulae given here include safety factors that are adequate for the majority of situations; however, where the loads are beyond the values considered here (see Section 3.2), larger margins of safety or specialist input are advisable. The formulae recommended in this manual, unless otherwise stated, were generally derived for conditions at the threshold of movement and therefore stable sizes are determined for very limited instability in the revetment, which should ensure that practically no damage to the revetment will occur. A recent design trend, however, has emerged that allows partial failure to occur. This can be suitable for situations where conservative hydraulic loadings are adopted, monitoring is frequent and/or a limited amount of damage is acceptable.

3.1. DESIGN PARAMETERS

There are a great number of parameters that generally need to be considered when selecting and designing a suitable revetment system. They can be grouped into the following categories.

- *Hydraulic.* Hydraulic parameters are essential for determining the stable sizes of revetments; some can be obtained from hydrological surveys and consultation of gauging records and others need to be calculated. Although most river engineers are aware of parameters such as the mean cross-sectional velocity and mean water level, others exist that may be crucial for determining the stability and extent of revetments. Among these are the local depth-averaged velocity, the velocity near the bed (see Section 2.3.1), waves (see Section 2.3.2), the level of turbulence in the flow (see Section 2.4.1), the velocity distribution in the cross-section (see Section 2.4.2), and in tidal reaches, the various water levels (mean, high and low), see Section 2.3.3. The values adopted for the hydraulic parameters will depend on the return period for which the revetment is to be designed. For example, the revetment of banks in flood defence schemes should be designed to give the same level of protection as the scheme itself, unless minor damage to the upper banks is accepted. Return periods can therefore range from a few years to several decades or more (particularly in urban areas).
- *Geotechnical.* Geotechnical parameters include the geometry of the existing channel (cross-sections and plan layout), soil characteristics and presence of scour or erosion (and historic instabilities).
- *Environmental.* Environmental parameters play an increasingly important role in the choice of strategy to limit or avoid erosion problems. Concern over preserving or enhancing the ecology of a given area (flora and fauna habitats) while creating aesthetically pleasant environments can be as important today as the consideration of engineering design parameters, such as the hydraulic stability of revetments. This factor also has implications for the construction methods to be adopted (see Chapter 6). Socially related issues also need to be addressed: likelihood of vandalism, recreation requirements (for example, fishing, rambling, water sports) as well as general needs of local land users (farmers, urban residents, industries).

- *Navigation.* Boat traffic (both recreational and for commercial use) generates waves (see Section 2.3.2) and requires minimum widths and water depths, as well as provision of moorings and maintenance of existing towpaths.
- *Construction.* Construction-related parameters include availability of materials and space, accessibility for machinery and workers, and under- or above-water construction. These issues are discussed in Chapter 6.
- *Maintenance.* The need for frequent and/or labour-intensive maintenance may impose an excessive cost on the scheme, as can the need to use grass-cutting machinery in places with difficult access. Maintenance is generally a heavier requirement for bioengineering and biotechnical solutions (see Chapter 7).
- *Costs.* The capital and maintenance costs of a scheme are parameters that will, in many cases dictate the choice of revetment. Cost-benefit analyses are often required to determine the most effective solutions. Although not easy to quantify, the environmental impact of the revetment should also be included in the analyses.
- *Statutory requirements.* These will vary with the country where the protection work is to be carried out, and are usually enacted by different bodies. Organisations that will normally need to be consulted are those with responsibility for the environment in general, for drainage of low areas, local authorities, organisations dealing with administration of navigable waterways, port and harbour authorities and government departments.

3.2. RANGE OF APPLICABILITY OF REVETMENTS

Before giving guidance on the types of revetment most suitable for each case, it is useful to define a few major characteristics that affect the overall behaviour of revetments. One of these concerns the very nature of the revetment, whether it is essentially formed by living plants (*bioengineering*), or by hard units (*structural* or engineered revetments) or by a combination of both types (*biotechnical* engineering). This classification was introduced in Section 1.2, where the scope of this book was defined and limited to structural and biotechnical revetments.

Another important characteristic is the flexibility of the revetment, defined as the ability to maintain good contact with the underlying soil during gradual settlement. According to this definition, a revetment such as a concrete lining is considered rigid. It has, in fact, little capacity to accommodate variations in the base soil without cracking, in spite of being able to follow bank contours effectively before it cures. Riprap is an example of a flexible revetment because of the ability of the individual stone units (each of them very rigid) to rotate next to its neighbours while remaining in close contact. Also, because riprap is formed by more than one layer of units, when erosion of the underlying soil occurs, the individual units can easily move to fill small gaps and holes. Flexible revetments are generally advisable for protection work, and particularly so for situations where soil instability is expected and where maintenance is to be kept to a minimum.

Permeability is also a relevant property of revetments. Revetments can range from almost totally impermeable (typical examples being concrete or fully-grouted stone) to various degrees of permeability. The importance of permeability is related to the ability of the revetment to transfer pressure forces in both directions, i.e. from the watercourse to the protected bed or bank and vice versa. This allows the release of pressures that can otherwise accumulate behind the revetment (see Section 2.1). However, if the bank material is impermeable, an impermeable revetment can be very stable (for example solid blocks on clay) because the external flow forces cannot generate any differential pressures across the layer. If some permeability is necessary (because the bank is permeable), then, even in extremely impermeable revetments, provision can be made for the release of these pressure forces (by means of weepholes) that may render the revetment suitable. Conditions where permeability is particularly relevant include rivers and channels subjected to tidal flows and waves (e.g. navigable reaches) and any other cases of variation in water levels and rapid drawdowns (as mentioned in Section 2.3.3).

The various kinds of hydraulic loading that ought to be considered in the design of revetments were described in some detail in Sections 2.3 and 2.4. As part of the first stages of design, it is generally advantageous to adopt a classification system that combines the primary loads (current attack and waves) and groups them in classes. These classes provide a simple way of assessing the severity of the attack and directing the engineer towards suitable types of revetment. Table 3.1 gives a recommended classification system, which is based, as are other systems, on currents and waves (considered separately). The various classes (light, moderate, heavy and very heavy loadings) are applicable to the flow velocities and wave heights that are likely to occur in rivers and canals. A fifth class, high turbulence, refers to the specific case of reaches downstream of hydraulic structures or bridge piers and other highly turbulent environments (see Section 2.4.1).

The range of applicability of the revetments considered in this book (plus bioengineered solutions) is summarised in Table 3.2. For description and detailed information on the various revetments, refer to Chapter 4. Table 3.2 is intended to provide a very general guide to the properties of each type of revetment and it

Table 3.1. Classification of hydraulic loading

Hydraulic loading		Classification
U*: (m/s)	$H^{\dagger}$: m	
≤ 1	$\leq 0{\cdot}15$	Light
1–2·5	$\leq 0{\cdot}50$	Moderate
2·5–4	$\leq 1{\cdot}0$	Heavy
4–7	$\leq 1{\cdot}0$	Very heavy
Downstream of hydraulic structures, around sharp bends, bridge piers, at transitions, etc.		High turbulence

*U—mean cross-sectional flow velocity.
†H—significant wave height of wind waves or boat wave height.

Table 3.2. Range of applicability of revetments

Hydraulic loading	Revetment types															
	Bio-engineering	Riprap	Block stone	Hand pitched stone	Grouted stone	Box gabions	Gabion mattresses	Sack gabions	Loose blocks	Cabled blocks	Bituminous materials	Filled sacks	Flexible form mats	Geomats and soil reinf. sys.	Concrete	Piling
Light	✓	✓	✓	✓	✓	✓	✓	✓	✓	✓	✓	✓	✓	✓	✓	✓
Moderate	✗	✓	✓	✓	✓	✓	✓	✓	✓	✓	✓	✗	✓	✓	✓	✓
Heavy	✗	✓	✓	✓	✓	✓	✓	✗	✓	✓	✓	✗	✓	✓	✓	✓
Very heavy	✗	✓	✓	✗	✓	✓	✓	✗	✗	✓	✓	✗	✓	✓	✓	✓
High turbulence	✗	✓	✗	✗	✓	✓	✓	✗	✗	✓	✓	✗	✗	✗	✓	✓
Bank slopes																
< 1*V*:1·5*H* (< 34°)	✓	✓	✓	✓	✓	✓	✓	✓	✓	✓	✓	✗	✓	✓	✓	✗
< 1*V*:1·5*H* (> 34°)	✓	✗	✓	✓	✓	✓	✗	✗	✗	✗	✗	✓	✗	✗	✓	✓
Near vertical	✓	✗	✓	✗	✗	✓	✗	✗	✗	✗	✗	✓	✗	✗	✓	✓

should be borne in mind that wide variability may exist between the different revetments, particularly among proprietary ones.

3.3. DESIGN STAGES

3.3.1. *Conceptual design*

The first stage of design usually involves the identification of the kind of revetment (or other type of strategy) that is most suitable for the case under analysis. It is essentially set at the conceptual level, hence its name, and requires very few, if any, quantitative calculations. What it does require, however, is a comprehensive appreciation of the various determining factors, which largely exceed the hydraulic parameters. The various phases of the conceptual design are shown in flowchart form in Figure 3.1; some of these are self-explanatory but others will be expanded further. This book, and particularly the section on conceptual design stage, is intended to be used in conjunction with the publication commissioned by the UK Environment Agency *Waterway bank protection: a guide to erosion assessment and management* (Morgan *et al.*, 1998), which was developed specifically for UK bank protection. It nevertheless provides important information for application in bed protection and overseas schemes.

Morgan *et al.* (1998) is a useful guide in the earlier phases of design, such as the assessment of the stability of the watercourse and the identification of the types and causes of bank failure. Among these, it is worth mentioning human and animal activity, river morphological processes and navigation. The designer is then asked to consider the implications of allowing the erosive process to continue or not, in view of the possible consequences for local residents, nearby structures and the environment. The definition of objectives for erosion control is carried out next, involving the consideration of the relevant design parameters described in Section 3.1. The accessibility of the site and other construction issues should also be studied at this stage. The above parameters will lead to the choice of strategy to be adopted (see Section 3.2):

- to allow natural adjustment of the channel
- to manage the erosion problem by practices such as fencing the banks to prevent overgrazing or
- to carry out engineering work.

Limiting flow velocities and wave heights, as well as qualitative information are given in the flowchart of Figure 3.1 to assist the engineer in the choice of revetment type. The value of velocity normally used at this stage is the average cross-sectional velocity corresponding to bank-full conditions. Bioengineering and management practices are not covered in this book and therefore, if these are the preferred solutions, the designer is recommended to consult other publications, such as Morgan *et al.* (1998) or Hemphill and Bramley (1989).

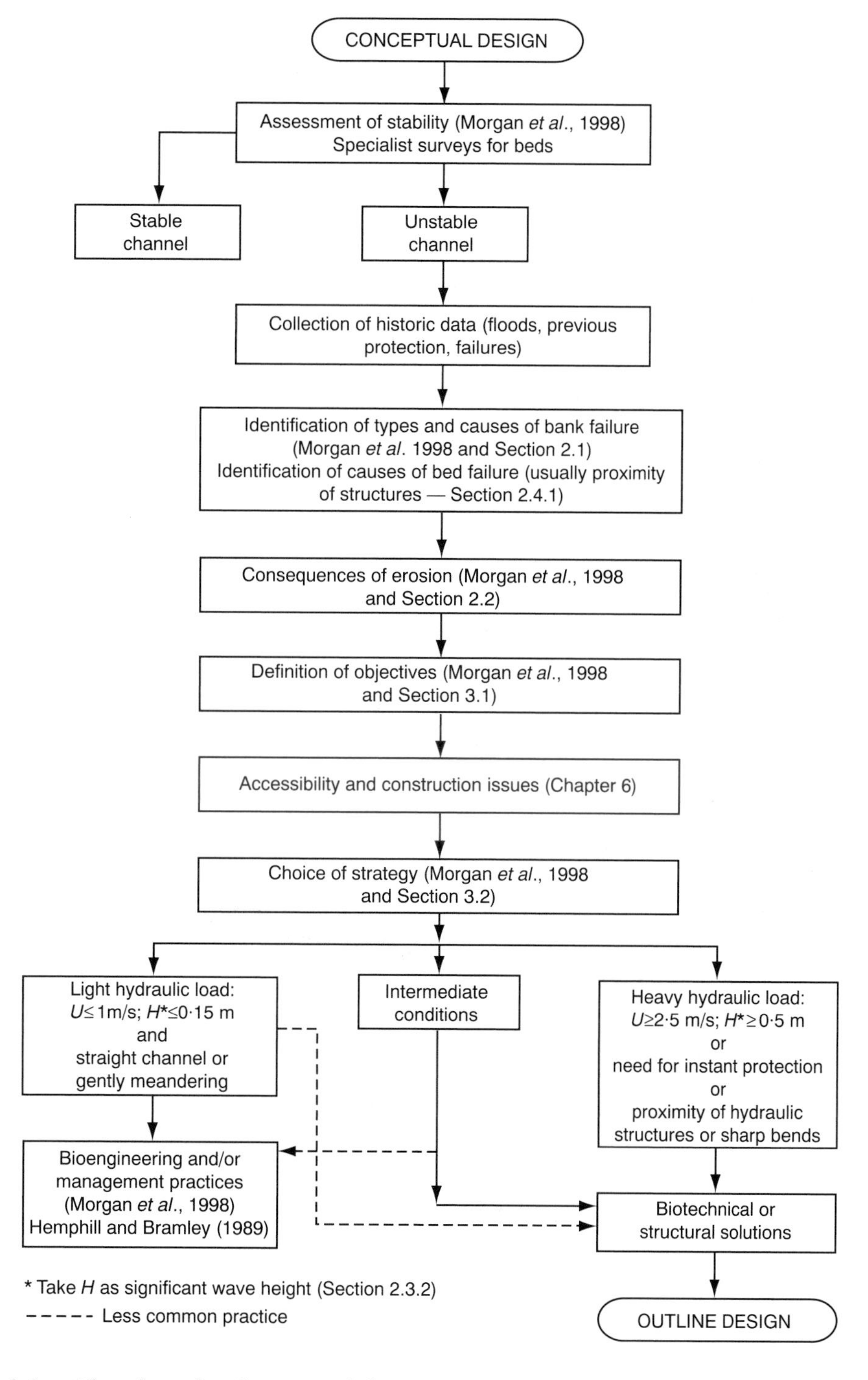

Figure 3.1. Flowchart for Conceptual design

3.3.2. *Outline design*

As mentioned in the previous section, the following stages of design regard solely biotechnical and structural solutions, that is, revetments incorporating some form of hard, engineered protection. Once one of these types has been selected in the conceptual design stage, the next phase is the outline, or preliminary design. The various steps in this stage are presented in the flowchart of Figure 3.2.

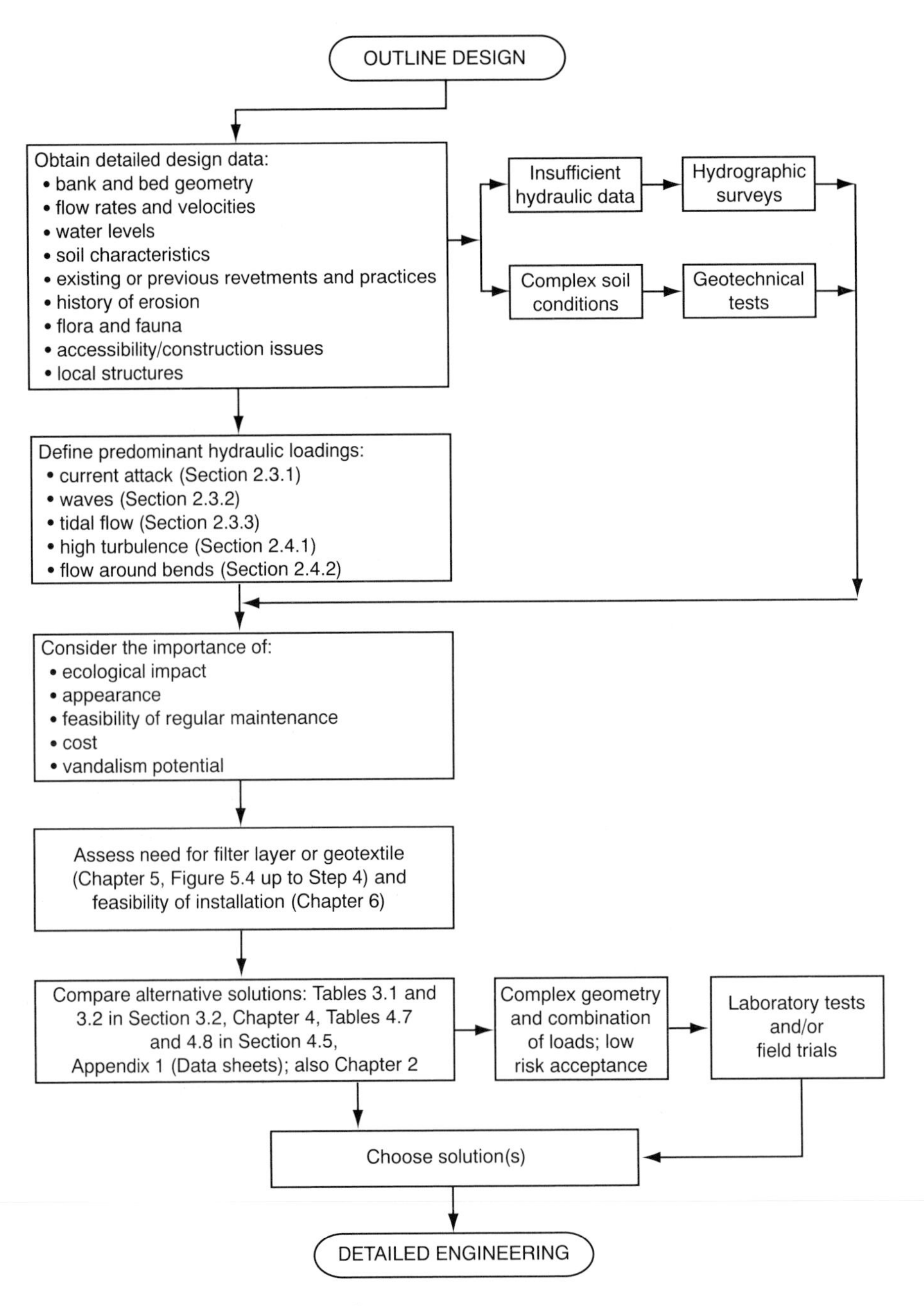

Figure 3.2. Flowchart for Outline design

Following the choice of type of strategy to pursue, the designer then needs to collect sufficient data to enable a technically supported decision on the best revetment type. The quantity and accuracy requirements for the data will depend, among other factors, on the importance and magnitude of the scheme, but have to be sufficient for the determination of the predominant hydraulic loads. Lack of hydraulic or geotechnical data can be overcome by commissioning hydrographic surveys and specific geotechnical tests.

The outline design proceeds with consideration of factors such as the ecological sensitivity of the site, the importance of cost and of attractiveness, the feasibility of carrying out frequent maintenance and the likelihood of damage by vandalism. Another important step is the assessment of the likely need for filter layers underneath the revetment and of any difficulties related to their construction or installation.

A comparison of alternative solutions will usually be possible at this stage. However, due to their complexity, certain design conditions may require laboratory tests or field trials to determine the suitability of some of the options under the specific site conditions. This may also be the case where a revetment is being considered as a possible solution for which independent or reliable design data is not available.

At the end of this stage, a single preferred type of revetment would normally be chosen for short river and canal stretches, whereas larger schemes are often found to need more than one type, since hydraulic and other conditions will vary along their length.

3.3.3. *Detailed engineering*

The last stage in the design procedure involves the detailed engineering of all the components of the chosen revetment (the cover layer, any granular or synthetic filters and the foundation layer) and the specification of any fixing requirements at the edges. Particularly important are the adequate design of bank toe protection and the detailing of transitions between the revetted reach and other existing protection or natural river banks and bed. Specification of construction methods, such as machinery required and timing of construction is also normally carried out at this stage. A detailed cost analysis listing capital and expected maintenance costs should also be produced for the whole design life of the scheme. In some cases, a cost-benefit analysis which would have started in the previous design stage, justifying the choice of revetment will now be completed. In this analysis an attempt should also be made, whenever possible, to quantify environmental benefits and costs.

The various steps of the detailed engineering design stage are presented in the flowchart of Figure 3.3.

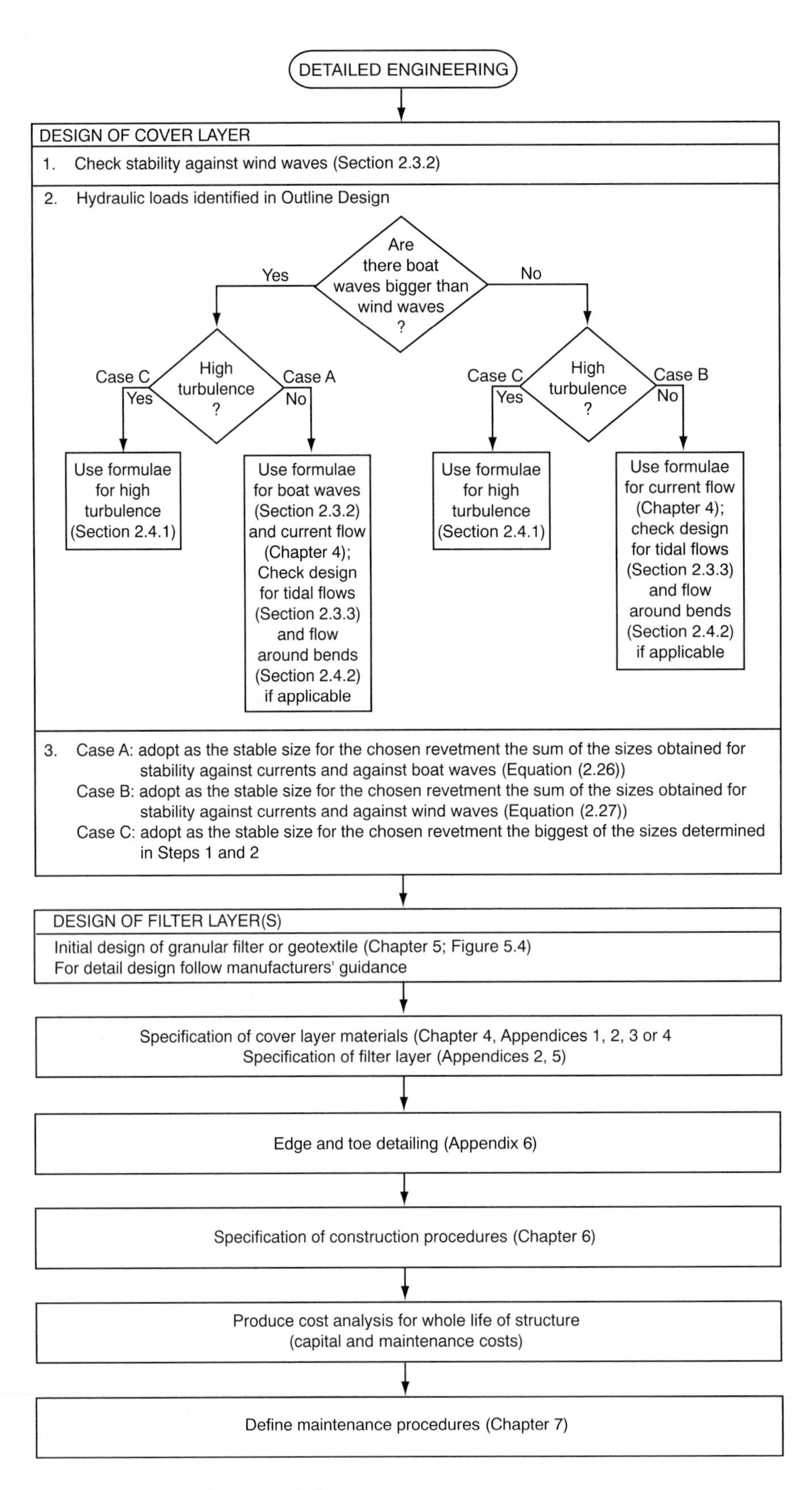

Figure 3.3. Flowchart for Detailed engineering

Types of revetment and design formulae

4. Types of revetment and design formulae

This chapter describes in detail common types of revetment system that are available to UK engineers, and is complemented by data sheets giving information on various proprietary systems (see Appendix 1). The list is quite extensive, but some less well known types may not have been included. The designer is therefore recommended to seek further information for situations requiring unusual or very specialist revetments. Although outside the scope of this book, it was decided to refer to piling, a system that is specifically adopted for vertical bank protection. This stemmed from its extensive use in river engineering, thanks to its very high resistance to hydraulic attack as well its ability to be combined with other, softer, types of protection.

Wherever possible, design formulae and general guidance are given for each type of revetment. These are suitable for design of straight river and channel reaches under current attack only; for other kinds of hydraulic loading (waves, tides, high turbulence) and for other river situations (sharp bends and vicinity of structures) the formulae and recommendations given in Sections 2.3.2 to 2.4.4 should be used. If information is lacking for certain types of revetment, the designer should consider commissioning laboratory tests to assess the stability of the revetment under particular flow conditions or conduct site trials.

4.1. ROCK

Due to its ability to resist heavy current and wave attack in a cost effective way, rock is and has been for many years one of the most popular materials used in the protection of river beds and banks. The use of rock covers a very wide range of applications from protection against the direct impact of flow (as a cover or armour layer, this latter term being more common in coastal engineering), to the construction of filter layers and regulating underlayers. Rock in revetments can be dumped randomly (e.g. riprap), carefully placed as loose units (e.g. block stone or hand pitched stone), contained within boxes (gabions and mattresses) or grouted by cementitious and bituminous mixes. Also, stone aggregate is obviously an important constituent in concrete revetments.

Rock has a number of properties that may be categorised as follows according to CUR (1995).

- Intrinsic properties.
  - *Density*. This is the ratio of dry mass to volume. Most rocks have densities in the range 2500 to 2700 kg/m^3 although the densities of some sedimentary rocks and basalt are 2000 kg/m^3 and 3100 kg/m^3, respectively. Rock density

is a good indication of strength.
The *relative density* is a property commonly quoted; it is the ratio between the stone density and the density of water. In freshwater rivers and channels the density of water can be taken as 1000 kg/m^3; in saline environments the density of water is normally taken as 1025 kg/m^3. For example, quartzite has a relative density of 2·6 in fresh water.

   - *Water absorption*. This is the mass of water absorbed per unit of dry rock mass. It is linked with in-service durability.
   - *Weathering*. This can be caused by temperature variations (particularly freeze/thaw) and by biological, chemical or mechanical attack.
   - *Fabric strength and presence of discontinuities*. These are important properties as they determine the rock's ability to resist breakage and damage abrasion.
   - *Colour*. The colour and general appearance of the rock can be important in situations where the rock must integrate well with the environment.

- Production-related properties.
   - *Block shape*. Rock used in river revetments can be elongated (particularly suitable for hand pitching and certain gabion applications), cuboid or irregular (these being either angular or rounded). The block shape affects the design because it influences the layer thickness achievable, the porosity and stability of the revetment.
   - *Block integrity*. The integrity of the stone units is affected by handling. Rough handling increases the likelihood of breakages and may result in a reduction of the mean stone size.
   - *Grading*. This is the size or mass distribution of rock, which is usually required for the assessment of hydraulic stability, for the design of filters and for the choice of equipment and construction methods. Since this is a very important property that is specified in all cases, it will be addressed again in detail later.

- Construction-induced properties (applicable essentially to riprap and gabions).
   - *Porosity*. This is the volume of voids per unit volume of rockfill. Although it depends on the level of compaction, common values to assume in design are between 15 and 40%. The lower limit corresponds to well-graded materials, as it is directly linked with the ability of smaller particles to fill the voids between the larger particles (interlocking).
   - *Internal friction*. This property is represented by the angle of internal friction (see Section 2.1.2) and mainly affects the stability of revetments on banks.

Because of its importance, rock grading is considered in more detail next.

Grading

It is advantageous from the viewpoints of production and quality control to think in terms of three different standard grading classes. One useful classification is shown in Table 4.1, which is based on Dutch Standards (note that the term 'weight' is commonly used to refer to mass).

The most convenient way of showing the mass or size distribution of rock is by means of graphs similar to those used for soils, depicting grading curves (see Section 2.1.1). The x-axis is a logarithmic scale of sizes (or weights) and the y-axis is a natural scale of percentages, so that the curve shows percentages of stone that are smaller (or lighter) than the corresponding sizes (or weights). For example, W_{50} is the weight of stone for which 50% of the total sample is formed by lighter particles. Knowledge of certain weight or size fractions is particularly important for revetment and filter design. The fractions considered are usually the W_{50} in the case of cover layers, and the D_{85} and D_{15} in the case of filters, and can determine the performance of a revetment system, as explained in Sections 4.1.1 and 5.3.1.

Grading curves can be classified according to the range of sizes they encompass, which is reflected in the steepness of the curve. CUR (1995) gives the classification reproduced in Table 4.2 to define in a systematic way a few terms commonly used in engineering rock applications.

As for other types of revetment, the choice of rock as a river bed and bank protection material depends on a number of factors, not all related to the capacity to perform well hydraulically during the design life of the scheme. Other considerations, described in general terms in Chapter 3, will influence the engineer's decision to a greater or lesser extent, depending partly on the country or region's regulations and environmental policy.

The decision to exclude stone from areas where it does not occur naturally (for example in the alluvial plains of tidal river reaches) is, in many ways, subjective and driven by aesthetic reasons. Some argue that stone is a natural product and will therefore always fit well into the landscape, especially after biological growth has had time to become established. Others will advise the use of stone only in

Table 4.1. Grading classes for rock

Grading class	Rock defined in terms of	Application
Heavy gradings	Weight (tonnes)	Armour layers in coastal works, individually handled stone (such as block stone) and cover layers
Light gradings	Weight (kg) or size (mm)	Armour and cover layers, underlayers and filters
Fine gradings	Size (mm)	Stone smaller than 200 mm (or 250 mm) obtained by production screening

Table 4.2. Grading definition

Grading	D_{85}/D_{15}	W_{85}/W_{15}	Observations
Narrow	1·2 to 1·5	1·7 to 3·4	Also called 'single graded' or 'uniform'
Wide	1·5 to 2·5	3·4 to 16	Riprap normally falls into this class
Very wide	2·5 to $>$5·0	16 to $>$125	Also known as 'quarry run'

upland areas, and are concerned about the over-exploitation of a limited natural resource. In the following sections indications are given of the suitability of the various types of stone used as revetments, but this necessarily involves a certain amount of subjective judgement.

4.1.1. *Riprap*

Riprap is the term used to describe loose quarry stone with a wide grading (D_{85}/D_{15} = 1·5 to 2·5 and W_{85}/W_{15} = 3·4 to 16) that is used for the protection of beds and banks against hydraulic forces (see Figure 4.1). Riprap revetments are formed by randomly placing layers of light grading stone (according to the Dutch classification given in Table 4.1) and sizes are typically greater than 200 mm or 250 mm. Riprap is specified both by weight (most frequently) and by size. Very large units (heavy gradings) above, say, 1000 kg are normally only required for marine works for protection against large waves, and form what is known as the 'armour layer'. In rivers this term is not commonly used; the term 'cover layer' is used in this manual to define the top layer of riprap or of any layer placed on top of other layers.

Riprap is one of the most versatile types of revetment as it can be specified to suit a very wide range of flow and soil conditions. Being formed by loose

Figure 4.1. Example of riprap protection on a river bank

individual units, it is quite flexible and can accommodate small ground movements or loss of some particles without collapse. Furthermore, because of the thickness of the cover layer, failure tends to occur gradually, allowing some time for repairs to be carried out. Although not universally accepted, it is thought that better stability is achieved with well graded riprap, where the smaller stone fraction partly fills the voids between the layer particles. The porosity is, in this case, considerably smaller than for uniform size stone (from 25% compared to almost 40%).

Stone of various shapes can be found in riprap revetments, from rounded to angular or elongated. These latter are generally considered less suitable because the elongated shape appears to confer lower stability; they are also likely to experience more difficulty in readjusting to new positions in the event of settlement or partial failure. Some failures of riprap revetment schemes have been attributed to the use of rounded stone. However, there is still some controversy about this issue since some research studies have not confirmed the above. Research carried out in the UK on riprap stability under turbulent flow conditions has also found that there is no appreciable reduction in stability for rounded stones when compared with angular ones (Escarameia and May, 1992). If possible, stones of blocky shape should be specified to avoid uncertainties in their performance.

It was mentioned above that riprap sizes are given either in terms of weight or dimension. Stone dimensions can sometimes be presented as the *side of the equivalent volume cube*, D_n, or the *diameter of the equivalent-volume sphere*, D_s. These can be related to the block weight (see earlier comment) W by the following relationships:

$$D_n = (W/\rho_s)^{1/3} \tag{4.1}$$

$$D_s = 1{\cdot}24(W/\rho_s)^{1/3} \tag{4.2}$$

where ρ_s is the density of the stone (typically in the range 2500 to 2700 kg/m^3).

It is also common to find riprap (particularly smaller sizes) defined as a result of sieve analyses, in terms of D_x, which is the size for which x% of the stone is smaller. Most stability formulae are based on D_{50} but some researchers have found that D_{30} is a better indicator of the nominal stable size. The following relationships are given to allow conversion between these sizes:

$$D_{n50} = 0{\cdot}84 \text{ to } 0{\cdot}91 D_{50} \tag{4.3}$$

$$D_s = 1{\cdot}13 D_{50} \tag{4.4}$$

$$D_{30} \approx 0{\cdot}70 D_{50} \text{ or more precisely } D_{30} = D_{50}\left(\frac{D_{15}}{D_{85}}\right)^{0{\cdot}32} \tag{4.5}$$

Table 4.3 summarises some of the major characteristics of riprap.

Table 4.3. Riprap characteristics

Grading	W_{85}/W_{15} = 3·4 to 16 Smooth, well graded curve
Shape	Ideally blocky (maximum dimension ≅ 3 × minimum dimension)
Angle of repose Angle of internal friction	Typically between 35° and 42° Typically between 40° and 45°
Specific gravity	Typically between 2·5 and 2·7
Nominal size:	Given in terms of D_x or W_x, which are the size or weight of stone for which x% of the stone is smaller
Revetment thickness	Not less than 2 × D_{n50} or 1 to 1·5 × maximum dimension
Porosity	Ranges between 25% (well graded riprap) and 40%

Design

Research on the stability of riprap under current attack has shown that the stable stone size is essentially dependent on:

- the velocity of the flow
- the flow conditions (such as the level of turbulence present and the degree of development of the velocity profile)
- the properties of the stone (namely its density relative to that of water)
- the place where the riprap is installed (whether on a bed or on a bank).

Many researchers have proposed formulae for design of riprap and designers are sometimes faced with the problem of choosing a suitable solution from a range of equally valid equations that give quite different results. These discrepancies may arise from an over-simplification of the parameters that affect riprap stability; over-simplified equations probably result from the desire not to overload the designer with variables that are sometimes difficult to quantify in practice.

Although a single equation should ideally be recommended for design, it is considered that, due to discrepancies between riprap formulae, this is not the best approach. Instead, three different equations were selected and are presented here. The designer will find that they may give different stone sizes for similar flow conditions but these should not be excessively disparate. It is recommended that individual judgement be used to select the solution for adoption. In some circumstances it may be advisable to use the average of the results, but when a conservative approach is thought to be best, then the bigger stone size should be adopted.

The equations presented next are valid only for the protection of straight river and channel stretches, and for normal levels of turbulence (see Section 2.4.1 otherwise). These equations are simplified forms of Equations (2.17), (2.19) and (2.22), which were presented in their full form in Section 2.4.1.

1. Escarameia and May (1992)

   A suitable safety factor is already included in the following equations, which are valid for bank slopes of $1V{:}2H$ or milder and normal river flow.

$$D_{n50} = 0{\cdot}050\left(\frac{U_d^2}{s-1}\right) \quad \text{for continuous revetments} \tag{4.6}$$

$$D_{n50} = 0{\cdot}064\left(\frac{U_d^2}{s-1}\right) \quad \text{at edges of revetments} \tag{4.7}$$

   where
   D_{n50} is the characteristic stone size

$$D_{n50} = \left(\frac{W_{50}}{\rho_s}\right)^{1/3} \tag{4.8}$$

   where
   W_{50} is the particle weight for which 50% of the total sample is lighter
   ρ_s is the stone density

   s is the relative stone density, defined as ρ_s/ρ. Usually s can be taken as 2·5 to 2·7 in freshwater applications — see Section 4.1
   U_d is the depth-averaged velocity of the flow. For the design of bank protection, U_d should refer to the toe of the bank. If values of U_d are not available, refer to Section 2.3.1 or use the mean cross-sectional velocity U, obtained by dividing the discharge by the cross-sectional area of the flow.

2. Pilarczyk (1990)

$$D_{n50} = \frac{\phi}{s-1} K_h K_s^{-1} \frac{U_d^2}{2g} \quad \text{for normal river flow} \tag{4.9}$$

   where
   D_{n50} is the characteristic stone size

   D_{n50} defined as in 1. above

   ϕ is a stability correction factor

   ϕ = 0·75 for continuous protection
   ϕ = 1·0 to 1·5 for edges and transitions (see note at end of 2.)

   s is the relative stone density, defined as ρ_s/ρ. Usually s can be taken as 2·5 to 2·7 in freshwater applications — see Section 4.1
   K_h is the depth factor

$$K_h = \frac{2}{[\log_{10}(12y/2D_{n50})]^2}$$ for fully developed velocity profiles, i.e. continuous revetments, away from edges or transitions

or

$$K_h = \left(\frac{D_{n50}}{y}\right)^{0.2}$$ for partially developed velocity profiles (see note at the end of 2)

K_s is the slope factor — see Section 2.4.1 for definition
y is the water depth. For bank protection take y as the depth at the toe of the bank
U_d is the depth-averaged velocity. For bank protection U_d should refer to the toe of the bank. If values of U_d are not available, refer to Section 2.3.1 or use the mean cross-sectional velocity U, obtained by dividing the discharge by the cross-sectional area of the flow
g is the acceleration due to gravity.

Note: values of ϕ greater than 1·0, used at edges and transitions and for partially developed velocity profiles, are more likely to be associated with increased levels of turbulence. Highly turbulent conditions are addressed in Section 2.4.1.

3. Maynord (1993)

Note that this equation is given in terms of D_{30}

$$D_{30} = S_f C_s y \left[\left(\frac{1}{s-1}\right)^{0.5} \frac{U_d}{\sqrt{(K_1 g y)}}\right]^{2.5}$$ for normal river flow (4.10)

where
D_{30} is the characteristic riprap size of which 30% is the finer by weight. Relationships between D_{30} and D_{n50} are given earlier in this section (Equations (4.3) to (4.5))
S_f is a safety factor (Maynord gives minimum value = 1·1 but a higher safety factor of 1·5 is suggested here for design) — see also Section 2.4.1
C_s is a stability coefficient

C_s = 0·3 for angular rock
C_s = 0·375 for rounded rock

y is the local water depth. For the design of bank protection take y at the toe of the bank
s is the relative density of stone, defined as ρ_s/ρ. Usually s can be taken as 2·5 to 2·7 in freshwater applications — see Section 4.1
U_d is the depth-averaged velocity. For bank protection U_d should refer to the toe of the bank. If values of U_d are not available, refer to Section 2.3.1 or use the mean cross-sectional velocity U, obtained by dividing the discharge by the

cross-sectional area of the flow
g is the acceleration due to gravity
K_l is a side slope correction factor

$$K_l = -0{\cdot}672 + 1{\cdot}492\cot(\alpha) - 0{\cdot}449\cot^2(\alpha) + 0{\cdot}045\cot^3(\alpha)$$

where α is the angle of the bank to the horizontal.
This equation is valid for side slopes between $1V{:}1{\cdot}25H$ and $1V{:}4H$.

The design of a riprap revetment is not confined to the determination of the stable stone size for the cover layer. It is equally important to check the need for a filter between the riprap and the underlying soil. This filter may consist of granular materials or of geotextiles, as explained in Chapter 5, which also gives recommendations for its design.

It is also worth mentioning that when riprap is dumped in flowing water some downstream displacement will normally occur. The displacement length L is a function of the flow velocity, the water depth and the stone size, as expressed by the following equation presented in Przedwojski *et al.* (1995):

$$L = 0{\cdot}25yU_dD^{-0{\cdot}5} \quad (4.11)$$

where
y is the local water depth
U_d is the local depth-averaged velocity
D is the nominal stone size.

Information on specifications for riprap in river applications is presented in Appendix 2.

Suitability

- Riprap can be designed to provide protection for both river bed and banks under heavy current and wave attack, including high levels of turbulence. Riprap is suitable for protection of banks with slopes up to $1V{:}1{\cdot}5H$ without provision of additional means of restraint.
- The ease of placement, which is carried out by machine and generally without any hand placing or compaction, makes riprap suitable for a number of situations, including underwater protection. However, this requires special care in flowing water to avoid loss of the smaller stone fractions. Installation of riprap blankets can be carried out from land (machine access needs to be available) or from barges (where land space is restricted).
- Being placed by machine, riprap is usually a cost-effective solution, particularly for relatively large schemes; due to its low maintenance requirements, it is also particularly appropriate for remote areas.
- Because of its flexibility, riprap is also a good material for protection against erosion at transitions between hydraulic structures and natural banks or other types of revetment.

4.1.2. *Block stone*

Although less common than riprap, block stone is nevertheless one of the materials used to protect river banks against heavy current attack. There is no precise definition of 'block stone' but the term is used in the UK to describe fairly square stone units which are placed by machine; each unit weighing typically over 1000 kg, which corresponds to stone dimensions above approximately 700 mm (see Figure 4.2). Block stone is carefully positioned in layers to form stepped banks or near-vertical walls. As shown in Figure 4.2, soil may be placed between and over the layers to help the establishment of vegetation; planting with local species is also sometimes carried out to help to bind the soil and improve the environmental quality of this type of revetment, but maintenance may prove difficult.

Three aspects are essential in order to ensure adequate stability of the revetment.

- The first is to build a good toe support, which can be usually achieved by placing at least one layer of large stone (generally greater than 1 m thick) below the original bed level.
- The second aspect is the provision of a filter behind the blocks of stone. A filter will practically always be necessary due to the large size of the stone units compared with the underlying soil, as explained in Chapter 5. In most cases, because of space limitations, synthetic filters (geotextiles) will be more advantageous than granular filters. An example of block stone construction is shown in Figure 4.3. In this figure the layer forming the toe has already been laid below the river bed level and the geotextile can still be seen, awaiting covering.

Figure 4.2. Example of a block stone protection

Figure 4.3. Block stone construction

- The third aspect regards the quality of the stone used. The stone needs to be hard and without cracks or discontinuities that would otherwise lead to breakages and consequent reduction of the revetment's ability to withstand the design level of flow attack.

Design

The design of block stone protection is in many cases based on considerations of past experience and good practice. In view of the lack of design formulae, it is particularly important to take into account the three aspects mentioned above. When the stone is used as a retaining wall, its stability should be checked for all the relevant modes of failure (see, for example, Hemphill and Bramley, 1989). For the design of the filter, it is recommended that the guidance given in Chapter 5 be used and that the other references mentioned there be consulted.

Suitability

- From aesthetic and environmental viewpoints, block stone is more suitable for the protection of banks of upland rivers.
- Hydraulically it can provide a stable protection under heavy flow attack but with relatively low levels of turbulence, since the porosity and flexibility of this type of revetment are quite low.
- Due to the possibility of building steep banks and forming retaining walls, block stone can be an attractive but generally more expensive alternative to box gabions or concrete wall units in situations where space is limited.

4.1.3. Hand pitched stone

This is a traditional form of bank protection that can withstand medium to heavy flow attack, and provides in some areas an aesthetically pleasant and neat revetment. The stone is placed by hand in dry conditions and the required effect can be achieved by choosing local stone of suitable colour, shape and size to match the surrounding environment. Compared with riprap, it involves very intensive labour; nevertheless, some savings may be made in the amount and size of rock required since hand pitched stone is placed in a single layer (see Figure 4.4). The main characteristics of this type of revetment can be summarised as follows.

- Usually single sized stone, sometimes with gravel or smaller stone wedged between the larger units. A bedding or regulating layer may or may not be necessary.
- Placed in a single layer, forming a much smoother surface than that of riprap.
- Close jointed to minimise erosion of underlying soil.
- Relatively inflexible due to close jointing and embedment into the soil; limited ability to resist settlement and soil deformation.
- Requires frequent inspection to check integrity of revetment, as loss of even a single unit may trigger collapse by erosion of the soil formation.

Design

Partly because of its historical use and declining applicability, only approximate design guidelines can be given for hand pitched stone revetments. These guidelines were drawn from comparisons of performance with riprap.

According to Hemphill and Bramley (1989), for the same flow conditions (including wave attack) the size required for pitched stone is only 85% of that

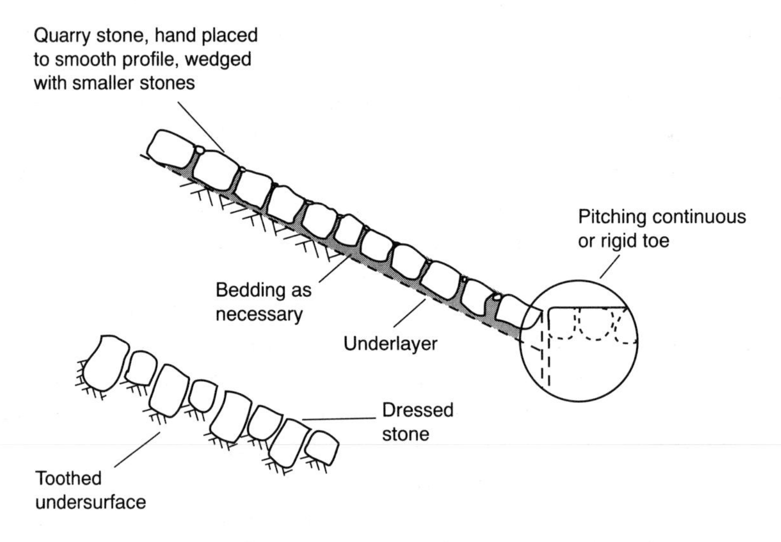

Figure 4.4. Typical examples of stone pitching (from Hemphill and Bramley, 1989)

required for riprap. The stable size can be as small as 70% for well keyed and embedded pitched stone. In the presence of waves, it is important to bear in mind that wave run-up can be 50% greater on hand pitched revetments than on riprap banks because of the smoothness of the revetment. As a rule of thumb, the height above still water level h can be estimated by: $h = 3H$, where H is the design wave height (see Section 2.3.2).

Suitability

- Hand pitched stone, because of the high labour costs involved in its construction and maintenance, is usually confined to small areas of revetment and to repairs of existing bank protection.
- It requires placement in dry conditions so, unless cofferdams are built or the channel has no flow, hand pitching only occurs on upper bank areas and rarely as bed protection. It is also essential to have a firm base soil into which the stone can be embedded; very permeable granular soils are, in principle, not suitable.

4.1.4. Grouted stone

In order to increase their stability and minimise the risk of loss of particles, stone revetments are sometimes grouted with cement or bitumen mortars. Compared with loose stone, grouting allows the use of smaller rock particles but the permeability of the revetment is necessarily reduced. When cement mortars are used, there is also a significant reduction in flexibility and the revetment becomes less able to adjust to ground settlements.

Grouting can be carried out on riprap and on hand pitched stone (Figure 4.5). In the case of riprap, bitumen is the material most commonly used; although various degrees of grouting are possible, effective solutions are usually produced when the bituminous mortar envelops the loose stone and leaves relatively large voids between rock particles. The degrees of grouting available are:

- surface grouting (which does not penetrate the whole thickness of the revetment and corresponds to about one third of the voids filled)
- various forms of pattern grouting (where only some of the surface area of the revetment is filled, between 50 to 80% of voids)
- full grouting (an impermeable type of revetment).

Detailed information on bitumen-bound revetments for river environments is presented in Section 4.4.1. Cement mortar is also used in conjunction with riprap, particularly to increase its stability at transitions with hydraulic structures or other types of revetment, and is best confined to small areas. Hand pitched stone is normally grouted with cement mortar where it is necessary to provide increased stability, such as near the confluence of streams or at inlet or outlet structures. The workability of the mortar generally needs to be increased by appropriate additives.

Figure 4.5. Example of grouted stone protection

Design

- *Bituminous grouted stone*
  It is known that reductions in the stable stone size are achieved by means of grouting but this effect has not been quantified accurately for the case of riprap protection in river applications. However, as a guide, Hemphill and Bramley (1989) indicate that the nominal stone size D_{50} required for wave attack, can be reduced by 10% with surface grouting; this value may increase to 40% for a pattern grouting corresponding to 60% of surface voids filled.
  Design guidance on types of bituminous bound revetment is given in Section 4.4.1.
- *Cement grouted stone*
  As mentioned above, this type of grout is more usually found in association with hand pitched stone; it confers rigidity and impermeability to the revetment, which need to be taken into account. In order to allow the release of pore pressures in granular soils, the revetment should include an underlayer and the grout should be omitted in a few places, so as to provide weepholes. Since this is a rigid type of revetment, it is important that the soil is firm (it can be compacted if necessary) and that adequate toe restraints are included in the design (e.g. a concrete toe beam or sheet piling).
  Several types of finish and methods of construction are available, depending on the level of appearance required. The cement grout can be poured over the stone or, for a more aesthetically pleasing effect, the grout can be confined to the gaps between the stones.
  Design guidance is scarce, probably due to the fact that cement grouting is adopted in many instances not so much to improve stability but rather to restrain the movement of individual stones.

Suitability

- Grouting is mainly used at edges of revetments and at transitions with hydraulic structures to ensure that the revetment can withstand the increased levels of turbulence resulting from changes in flow direction.
- Grouting can also serve a more aesthetic purpose in urban areas as well as being a deterrent against theft and vandalism.

4.2. GABIONS

Gabion is a generic name given to types of revetment that are usually formed by stone contained in wire mesh. These wire containers can vary considerably in shape, from cuboid and rectangular shapes with relatively small thicknesses, to sausage-like shapes. The first type is generally known as box gabions, the second as gabion mattresses and the third by various names such as sack gabions, rock rolls, tubular or sausage gabions (see Figure 4.6).

The objective behind the development of gabions was to achieve an increase in stability of stone, when compared with loose stone, by restraining its movement. This means that for the same flow conditions, smaller stone sizes can be used, which can be a major advantage for regions with a scarcity of large stone. The flexibility of the mesh boxes allows them to deform when subjected to current or wave forces without failing, while the rockfill is contained by the mesh. Many of the advantages of riprap are also present in gabions, permeability and flexibility being the major ones.

Due to their high permeability, rock gabions will not generally induce high differential pressures behind the revetment, and may not always require filters over the natural soil if this is easily drained. However, geotextiles are sometimes used to avoid loss of fines from the bank or bed. Being quite flexible, gabions will adapt well to small settlements without experiencing collapse; as mentioned above, the boxes can also suffer some deformation caused by internal movement of the fill.

Most rock gabion boxes are formed by wire or polymer mesh. Two types of wire mesh can be found:

- woven, which provides more flexibility to the mesh box
- welded, a more rigid alternative which is considered by some to be easier to fill.

In both cases the wire can be galvanised and coated with PVC in order to prevent damage by corrosion. Corrosion can be a real threat in saline and peaty environments, as well as near to sewer and industrial outfalls. Galvanised wire is standard for gabions from most reputable manufacturers. Typical mesh sizes vary between 60 and 80 mm and are obviously determined by the smallest size of the infill.

Although the great majority of box gabions are made of wire or polymer mesh with stone as their fill material, a few types are also available for the containment of soil, sand or other small size aggregates. In these cases the boxes are either lined with geotextiles to prevent the escape of the small infill particles or are made from fine polymer mesh.

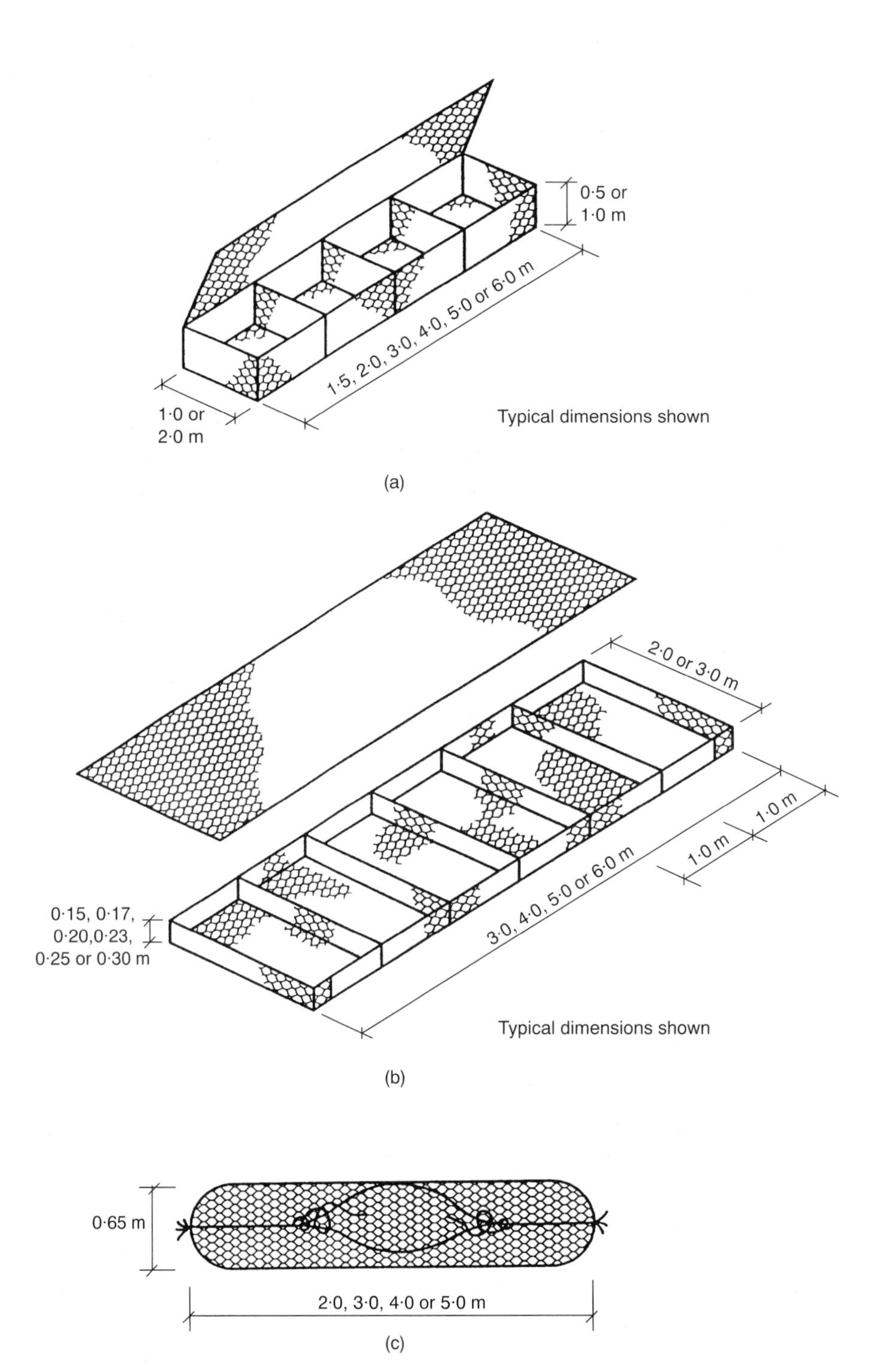

Figure 4.6. Examples of gabion types (adapted from Hemphill and Bramley, 1989): (a) box gabion; (b) gabion mattress; (c) sack gabion

After being delivered in flat packs, the gabion boxes are usually assembled on site, and then filled with stone or other materials. Overfilling is generally recommended to allow for settlement. The filling can be done by simple machinery or by hand, for example when local blocky stone is specified to achieve a traditional dry-wall look. Pre-filling and positioning underwater can also be carried out with the aid of lifting hooks, frames or pontoons.

Gabions are susceptible to two main types of damage that can reduce their serviceable life:

- damage by abrasion
- vandalism.

Abrasion of the wire mesh can result from the external action of sediment-laden flow. Continuous attack, particularly in highly turbulent conditions (e.g. hydraulic jumps), causes weakening of the mesh and can eventually lead to the loss of infill. External abrasion and susceptibility to corrosion do not constitute such a severe problem as in the past since galvanised wire and additional PVC coating are available readily and are actually specified for most applications. Abrasion of the fill material can also occur as a consequence of the internal movement of stones; these can break into particles smaller than the mesh size, which will then be washed away. In areas very prone to vandalism, cutting of the wire boxes and the theft of the rockfill can jeopardise the stability of the revetment, and therefore it is advisable in these areas to choose alternative types of revetment.

To conclude this introduction on gabions, an important point should be stressed. Gabions can be fabricated by almost anyone, but the designer should only specify proprietary products that have been the object of comprehensive research on their mechanical and hydraulic properties, or have a well documented, successful history. Data sheets for various proprietary gabions can be found in Appendix 1.

4.2.1. Box gabions

Box gabions are cage-like containers that are usually filled with stone. They are essentially cuboid in shape (see Figure 4.6). Transverse diaphragms are often introduced to limit the available space for movement of stones inside the boxes when subjected to heavy attack. They also prevent excessive bulging of the box during filling.

Different types of box gabions are available, varying in the type of mesh (wire, polymer or membrane), the method of wire mesh fabrication (woven or welded) and the nature of the fill material.

Soil filled gabions are mainly employed for a soil retaining function, in many cases fulfilling at the same time an aesthetic function because they can easily be part-filled with soil to sustain vegetation. Due to their bulkiness, easy handling and low cost, they are also used as embankment fills in flood protection schemes.

Rock filled gabions can also be designed specifically to provide an attractive vegetation cover (for example by including a turf mat or willow logs on the gabion lid) as illustrated in Section 4.2.4. Furthermore, they can have the necessary strength to be used as bed protection downstream of hydraulic structures.

Rock can be replaced by other filling materials as long as they are durable, heavy, chemically stable, easily obtainable and therefore of low cost. The following are some of the most common materials: concrete, bricks or other ceramic materials, certain industrial by-products (such as steel slags; copper and lead slags should be avoided due to the high concentrations of heavy metals that they contain).

Design

As mentioned above, box gabions can be used in the following two major types of application:

- as a revetment, particularly for protection of river and canal beds
- as a soil retaining structure (or gravity wall) — see Figure 4.7.

As a revetment, the use of box gabions to protect river banks is not very common because their square shape makes them less suitable unless the banks are very steep. Mattresses are generally preferred since they give a much smoother appearance to the banks and require less volume of filling materials. The design of box gabions, both as revetments and retaining walls, can in many instances be entrusted to the manufacturers if they can provide technical advice based on

Figure 4.7. Example of bank protection with box gabions

sound research. However, either because not all manufacturers invest in comprehensive research programmes, or because the designer needs to assess the validity of the manufacturers' suggestion, some design recommendations are given next.

It is likely that gabion boxes used for protection of river beds will be placed in areas of high turbulence intensity, such as downstream of culverts, weirs, outfalls or at the confluence between channels. Therefore, when designing box gabions as bed revetments it is suggested that the designer refer to Chapter 2 and use Equations (2.17), (2.19) or (2.22) to determine the required stone size. The sizes of box and, to some extent, of mesh are generally predetermined. In Equation (2.17), which was developed for 300 mm thick gabion mattresses, the value of coefficient C to use is $C = 12{\cdot}3\ TI - 1{\cdot}65$ (valid for $TI \geq 0{\cdot}12$, where TI is the turbulence intensity); values of TI can be found in Table 2.6 for various flow conditions. The various factors for use in Equation (2.19) can be found in Table 2.8 in the column referring to box gabions.

When designing box gabions as gravity walls, the stability of the wall should be assessed for all the relevant modes of failure (see, for example Hemphill and Bramley, 1989). Because they can accommodate small settlements of the foundation, these walls are generally built with a small angle to the vertical (of about 6°) to prevent them from tilting forwards in the long term. Being permeable structures, gabion walls do not generate high back pressures and therefore may not require the introduction of a filter between the bank and the wall if the base soil is free draining. The inclusion of filters may be advisable, however, to prevent loss of fines (see Chapter 5).

Suitability

- Box gabions are suited for the protection of river beds and banks subjected to heavy current attack. They can be stable at flow velocities of the order of 5 to 6 m/s (or more, depending on quality of construction and assembly).
- Since they require smaller stone sizes than riprap for the same flow conditions, they are a good alternative for regions where large stone sizes are difficult (and expensive) to obtain.
- They also have a long design life of several decades, particularly if the mesh is PVC coated or made of polymers.
- The stark appearance of box gabions can, to some extent, be 'disguised' by vegetation growth; but their cuboid shape will always appear somewhat unnatural, unless they are covered by a smooth layer of grassed topsoil (see Section 4.2.4). However, they can provide a reasonably good habitat for small animals and plants.
- In areas with a great risk of vandalism, box gabions may require fairly frequent maintenance to replace cut wire and stolen stones.

4.2.2. Gabion mattresses

Gabion mattresses are essentially rock-filled wire mesh boxes which have a large surface area compared to their thickness (see Figure 4.8). Diaphragms or internal transverse baffles are introduced to restrain the movement of the fill inside the boxes, which will tend to occur near limiting flow conditions. Most proprietary systems available in the UK consist of galvanised woven or welded wire meshes. The wire meshes are usually PVC coated for harsh environments, but polyethylene meshes are also manufactured for use in contaminated soil or water. By far the most common fill is stone; other possible materials include concrete, bricks and suitable industrial by-products. Under very severe hydraulic loading or where vandalism is a real concern, the stone fill can be grouted with cement mortar or bitumen mixtures. The use of cement grout should be limited to small areas (such as near bank toes, transitions and edges) as it renders the revetment rigid.

Schemes combining mattresses and box gabions are commonly found, as is the increased effort to help the establishment of vegetation on these systems. Examples of this trend towards more environmentally friendly solutions are presented in Section 4.2.4.

Design

The design of gabion mattresses consists primarily of the calculation of the stable stone (or particle) size to be contained by the mesh. The mesh openings are usually chosen later from the available range so that loss of stone will not occur. The necessary size of stone fill should then determine the minimum thickness of the mattress (1·5 times the nominal stone size is generally accepted).

Figure 4.8. Example of bank protection with gabion mattresses (courtesy of Maccaferri Ltd)

Manufacturers usually have a range of thicknesses to offer and the major manufacturers can also suggest maximum flow velocities for which the various thicknesses of mattresses are stable.

Although manufacturers can give very valuable information and design guidance, the following equation, from Escarameia (1995) and developed at HR Wallingford, is suggested here for the design of gabion mattresses under current attack in normal turbulence conditions. This is a simplified form of Equation (2.17) for the specific case of normal turbulence, which was presented in Section 2.4.1 in its full form:

$$D_n = 0{\cdot}007\frac{U_d^2}{s-1} \quad \text{for side slopes of 1V : 2H or milder and normal river flow} \qquad (4.12)$$

where

D_n is the characteristic size of stone

U_d is the depth-averaged velocity of the flow. For bank protection take U_d at the toe of the bank. If values of U_d are not available, refer to Section 2.3.1 or use the mean cross-sectional velocity U defined as the discharge divided by the cross-sectional area of the flow

s is the relative density of stone, defined as ρ_s/ρ. In freshwater applications, s will normally be between 2·5 and 2·7 (see Section 4.1).

It should be noted that the above equation was developed from laboratory tests on gabion mattresses 300 mm in thickness.

Suitability

- Gabion mattresses can have a very long serviceable life (several decades or more), require little or no maintenance and can withstand heavy current attack when placed on banks or beds.
- They can be placed under water as well as in the dry; large extents can be protected cost-effectively and later develop an adequate natural or induced vegetation cover.
- Gabion mattresses are particularly suited for the revetment of large areas, even in remote regions (since maintenance requirements are low), and for the protection of river beds and banks in the vicinity of hydraulic structures.
- In zones exposed to very high risk of vandalism, other types of revetment may be more suitable.
- As a very general guide for the assessment of alternative revetment solutions, the following table presents the limiting flow velocities for mattresses of two common thicknesses (note that the range of velocity values reflects differences in the alleged stability of the various proprietary products):

Mattress thickness: mm	Limiting velocity: m/s
150	< 2–3·5
300	4–5·5

4.2.3. *Sack gabions/rock rolls*

Sack gabions (also known as rock rolls, tubular gabions and sausage gabions) are cylindrical mesh containers filled with heavyweight aggregate such as rock, crushed concrete or bricks (see Figure 4.6). Flexibility is one of the major requirements for the mesh, since the rolls often need to be able to adjust to irregular contours. The mesh can be made of galvanised wire (coated with PVC in corrosive environments) or nylon. The ability to follow contours is, in fact, one of the main properties of sack gabions: their shape is ideal for filling scour holes and for protecting the toe of river banks.

The most common practice is to position the rolls along the toe, parallel to the main stream direction. In some applications the rolls are stacked to form steep banks or near vertical walls, and care is needed to ensure that the rolls are securely tied together. Sack gabions are pre-filled and placed in position by lifting cranes both in dry and underwater conditions. Less frequently employed than a longitudinal placement is the installation of gabions side by side down the slope of a bank. In this situation, the gabions are part-filled before placement or totally filled in situ, and are generally longer than the gabions used to fill scour holes, which are only 2 to 3 m long.

Biodegradable fibre rolls filled with soil, and sometimes incorporating plant bulbs and rhizomes, are a product related in shape to sack gabions. Their use is, however, quite distinct: they are placed near or above the water-line so that the plants can survive and are not suitable on their own for protection against heavy current attack. Being classified as bioengineering, fibre rolls are outside the scope of this book, but because they are increasingly being used in conjunction with structural revetments, further information is given in the data sheets of Appendix 1 and in Section 4.2.4.

Design

Very limited information is available on the design of sack gabions either in the literature or from manufacturers. The latter can, however, provide technical guidance to the designer and recommend suitable installation practices based on past experience.

When used to protect the toe of a bank, it is important to fix sack gabions securely by means of timber stakes or anchors driven into the slope. Good fixing is also important where several layers of sack gabions are used to form a retaining wall. In this case the design should include the assessment of all the possible modes of failure, as described for example in Hemphill and Bramley (1989).

Suitability

- Sack gabions are suitable for the filling of existing scour holes near river and canal banks, for the protection of bank toes under heavy current attack (flow velocities of the order of 2·5 m/s) and also for the formation of gravity walls. A less common application is the protection of banks with long sack gabions placed side by side transversely to the streamwise direction.

- As with other types of gabion, the size of stone fill is smaller than that of riprap in equivalent flow conditions and therefore these systems are suitable for regions where large stone is scarce.

4.2.4. *Composite types*

Gabions in all their multiple forms (boxes, mattresses, sacks) have very good characteristics from the hydraulic viewpoint: they are permeable, flexible and can provide long-term protection for river beds and banks in severe flow conditions. However, there is some concern that they do not always satisfy conservation and landscaping requirements. One of the major concerns is the wire or polymer mesh, which is undoubtedly an artificial material. The bulky shape of box gabions can also appear very unnatural in a rural environment. However, the high porosity of the gabion fill can provide adequate habitats for some species of flora and fauna. Even with little or no effort, grasses and small bushes have been found to establish successfully on gabions, almost totally covering their surface within a few years of construction (see, for example, Figure 4.9).

Positive action by gabion manufacturers and designers has led to the development and application of composite solutions, combining gabions with bioengineering solutions or with specially designed products, such as turfed mat lids. Examples of these solutions can be found in Figures 4.10 and 4.11. Specific information on choice of vegetation suitable for the UK can be found, for example, in Morgan *et al.* (1998).

Figure 4.9. Example of good vegetation cover over gabion protection (courtesy of Maccaferri Ltd)

Figure 4.10. Example of composite revetment consisting of gabion mattresses, box gabions and vegetation cover (courtesy of Maccaferri Ltd)

Figure 4.11. Example of combination of bioengineering and structural revetments (courtesy of Maccaferri Ltd)

4.3 BLOCK REVETMENTS

Pre-cast concrete blocks are commonly used to form revetments in rivers and channels. They are particularly suitable for regions where stone of moderate to big size is difficult or too costly to obtain. Concrete blocks for river and channel protection are generally factory-made and transported to site; in large schemes, however, it may prove more economical to install an in-situ plant.

Being mostly proprietary, there are a large number of different types and sizes of blocks available, but they fall into two broad categories:

- individual loose blocks
- connected blocks, usually by cables but also by means of an underlying geotextile.

Loose and cabled blocks are described in detail in Sections 4.3.1 and 4.3.2 and in Appendix 1, where information is given on the most common types available in the UK.

Although this section is devoted to single-layer block mattresses, protection of river beds and banks can also be achieved by dumping of pre-cast concrete blocks, in a similar way to riprap. In this case, the blocks are usually cuboid or tetrahedral in shape, and their stability can be estimated roughly using the equations recommended for riprap with allowance for the relative density of concrete (see Section 4.1.1). This type of revetment has been successfully applied in large training schemes, but experience has shown that, due to their weight, these blocks require coarse grained bed materials to minimise the risk of sinking into the soil base (Przedwojski *et al.*, 1995). Typical sizes of these blocks range from 0·3 m to 0·6 m (side dimension). In regions where stone is very scarce, sand–cement blocks and concrete blocks using broken clay bricks as aggregate have been found to perform well (Perkins, 1994).

Most blocks have plan shapes that create some degree of interlocking with the surrounding blocks, and a few proprietary systems are designed to rely heavily on this feature. Another characteristic, which is common to many patented blocks, is a slight upward taper. This feature is introduced to facilitate articulation of the joints and to allow the filling of interstices with sand, gravel or pea shingle to lock the units in place. However, the erosion protection given by these products when a good contact is established with the underlying soil is due essentially to their weight per unit area. For the same surface area, the thicker the blocks the more stable the revetment will, in principle, be.

Since these revetments consist of a single layer of blocks, it is very important to ensure the stability of each and every block under the design flow and for the geotechnical conditions of the site. In effect, the movement of one block, i.e. its lifting from the base material, will expose the base to the attack of the flow, and erosion may occur. Once the block revetment is undermined, collapse of the system is likely. This scenario is quite different from what happens in riprap revetments, where the gap left by the movement of a stone can generally soon be filled by another stone, thereby precluding sudden collapse. On the other hand, concrete block revetments present a much smoother surface than riprap and therefore drag forces are not as easily exerted on blocks as they are on riprap stones.

4.3.1. Loose and/or interlocking blocks

This category comprises revetments that are formed by the juxtaposition of blocks, ranging from a limited degree of connection with surrounding units to revetments where the blocks mobilise strong frictional bonds with their neighbours (see Figure 4.12). Both types consist of relatively small individual units, which can be, and normally are, placed by hand (although some

Figure 4.12. Example of bank protection with loose interlocking blocks (courtesy of MMG)

manufacturers have developed special machinery for their systems). This characteristic renders them ideal for the protection of small areas with difficult access for machinery. On the other hand, placement under water is only possible in about 0·5 m water depth and clear water conditions.

Some of the types considered in this category were originally designed, and are primarily used, for the reinforcement of car parks or lightly trafficked areas. However, they can be successfully used for erosion protection of channels and rivers, and are particularly suited for the revetment of berms.

Design

The stability of concrete blocks under current attack and in the absence of the destabilising actions of high turbulence, waves and other differential loads, is determined primarily by the flow velocity, the density of the concrete and the thickness of the blocks. This is true provided there is a good bond with the base layer and the blocks are laid on gentle slopes in such a way as to minimise protrusions of individual units of more than a few millimetres. Under these conditions, it is recommended that the following design equations be used, which already include adequate factors of safety and are valid for slopes of $1V$:$2{\cdot}5H$ or milder and for normal river flow:

$$D_{\rm n} = 0{\cdot}037U_{\rm d}^2/(s-1) \quad \text{for continuous protection} \qquad (4.13)$$

$$D_n = 0{\cdot}048 U_d^2/(s-1) \quad \text{at edges of revetments} \qquad (4.14)$$

where

D_n is the block thickness

s is the relative density of concrete, defined as ρ_c/ρ, where ρ_c is the density of concrete and ρ is the density of water. Usually s = 2·3 in freshwater applications

U_d is the depth-averaged velocity of the flow (or mean flow velocity). For bank protection U_d refers to the toe of the bank. If values of U_d are impossible to obtain, refer to Section 2.3.1 or use the mean cross-sectional velocity U obtained by dividing the discharge by the cross-sectional flow area.

The above equations are valid only for situations where bank or bed protection is needed in straight stretches of rivers or channels, and where the levels of turbulence are considered normal (see Section 2.4.1). This equation is a simplified form of Equation (2.17) using the relevant value of C from Table 2.7 presented in Section 2.4.1, which was derived from laboratory tests of loose blocks on a gravel bed (Escarameia, 1995). The reader should refer to Section 2.4.1 for flow conditions outside those listed above; when considering the placement of blocks on slopes steeper than $1V$:2·5H or the use of blocks with characteristics substantially different from those listed in Appendix 1, it is advisable to carry out laboratory tests to assess the stability. There are other aspects, besides the determination of the block thickness, that are important in achieving good design. One of these is to specify good drainage of the base soil (by means of a granular filter or a geotextile—see Chapter 5). Adequate restraint to the movement of the blocks down bank slopes, for example by means of concrete toe beams, is another good design procedure. Additional friction between blocks is usually generated by blinding (i.e. filling the interstices of) the blocks with pea shingle or gravel, although this is not (and should not be) taken into account in design.

Suitability

- Loose and interlocking blocks are primarily suited for the revetment of channel and river banks where access for machinery is limited or difficult.
- Being fairly labour intensive, they provide an economic alternative to other types of revetment where small to medium sized areas need to be protected against erosion. Because of the small size of the units, difficult contours such as bends and approaches to bridges, can be negotiated with relative ease.
- As for other loose unit revetments, attention should be given to the risk of vandalism or theft.
- The stability of concrete blocks depends essentially on their weight per unit area. Most manufacturers offer a wide range of blocks that vary not only in the ratio of thickness/surface area, but also in their shape, in the presence or absence of openings, and even in surface texture. This variability and the lack of comprehensive research on the stability of the many proprietary blocks

available makes it difficult to establish limits for application of these blocks. Nevertheless, research on grass reinforced channels (CIRIA, 1987) and laboratory studies carried out by HR Wallingford (Escarameia, 1995) on stability of loose and interlocking concrete blocks, allowed the derivation of the following guidelines:

Block thickness: mm	Limiting flow velocity: m/s
75 and 80	1·4
90	1·5
100	1·6
100 grassed	4·0
150	1·9
175	2·1

The above figures are valid for straight river flow, far from any hydraulic structures and with negligible wave and tidal effects. Note that the enhanced stability of grassed 100 mm thick blocks was found in conditions that were somewhat diverse from normal river flow: the block revetment was placed on a slope and subjected to overflow during short periods of 1 to 2 hours.

4.3.2. Linked blocks

Blocks can be linked together to form mattresses, which have the following advantages over loose blocks: reduced risk of progressive collapse of the revetment, feasibility of underwater placement and the covering of extensive areas with a small number of operations. The most common way of connecting blocks is by means of steel or synthetic cables running through purpose-made holes in the blocks. The panels thus formed are assembled in various sizes, most commonly from 7 to 30 m^2, but other sizes are achievable.

Cabled systems are usually installed on banks, with the cables running in the direction normal to the flow (see Figure 4.13). This results from the fact that the cables mainly serve the functional purposes of tying and of facilitating the fixing of the revetment at the top of the bank. However, cross-cabling is also available from some manufacturers for use where additional restraint is considered necessary.

An alternative to cabling is to bond or fix the blocks to a geotextile membrane of sufficient strength to withstand the lifting forces during installation. This system allows the whole revetment (i.e. the blocks and underlayer) to be placed under water, while minimising the difficulties and uncertainties of placing lightweight membranes under water (which would generally be required underneath cabled block mats). As for cable-tied mats it is important, however, to prepare the base formation well so that good contact exists between the revetment and the underlying soil.

Many manufacturers include both loose and cabled blocks in their range of products, the main difference being the presence of ducts for the cables. However, those that concentrate on cabled blocks tend to provide blocks with larger weights per unit area, which are designed to be stable under wave as well as current attack. These blocks are typically heavier than 200 kg/m^2 and can reach

Figure 4.13. Installation of cabled block mattresses (courtesy of MMG)

500 kg/m^2 with thicknesses of up to 225 mm. In small to medium sized non-navigable rivers, weights per unit area of the order of 140 kg/m^2 have been adopted with success; nevertheless, this does not preclude the need for careful design.

Design

As mentioned in Section 4.3.1, the stability of concrete blocks under normal river flow (see Section 2.3) depends on three major factors, as long as good contact is achieved with the soil base: the thickness of the blocks, the density of the concrete and the flow velocity. When compared with loose block systems, cable-tied blocks have the additional constraint of the cables which not only improves the fixing of the mattresses at the edges, but also slows down the process of any failure. This applies also to geotextile bonded block mats. In both systems an upward movement of the blocks is needed before the dead weight of the surrounding blocks or the restraining force of the cables is mobilised. Since this implies the local loss of contact with the underlayer, it is generally accepted that this property should be regarded as a reserve capacity of the system and not be assumed in design.

The work by Pilarczyk (1990) is recommended here for the design of concrete block mattresses under current attack in the absence of significant levels of turbulence or waves. The following equation is a simplified version of Equation (2.19) presented in Section 2.4.1, for the specific conditions listed below.

- The concrete mattress provides continuous protection (i.e. this applies to medium to long reaches of rivers and channels, far from transitions and edges).
- The non-dimensional critical shear stress, i.e. the Shields parameter, has a value of 0·07.
- The turbulence level in the river or channel is normal for straight rivers (see definitions in Section 2.4.1).

The thickness of the blocks can therefore be determined by the following equation. This equation involves an iterative procedure, i.e. an initial estimate of the block thickness needs to be assumed on the right-hand side:

$$D = \frac{0{\cdot}026U_d^2}{(1-n)(s-1)[\log(12y/D)]^2 K_s} \tag{4.15}$$

where

D is the block thickness
n is the porosity of the revetment, which can be obtained from the open area of the blocks (for example, for 25% open area, n = 0·25)
s is the relative density of concrete, defined as ρ_c/ρ, where ρ_c is the density of concrete and ρ is the density of water. Usually s = 2·3 in freshwater applications
U_d is the depth-averaged velocity of the flow (or mean velocity). For bank protection, U_d refers to the toe of the bank. If values of U_d are impossible to obtain, refer to Section 2.3.1 or use the mean cross-sectional velocity U obtained by dividing the discharge by the cross-sectional area of the flow
y is the water depth. For bank protection y is the water depth at the toe of the bank.
K_s is the slope factor (see CUR,1995), which is defined as the combination of two slope terms: a side slope term (k_d) and a longitudinal slope term (k_l) in the direction of the current:

$$K_s = k_d k_l \tag{4.16}$$

where

$$k_d = \cos\alpha\sqrt{\left[1-\left(\frac{\tan\alpha}{\tan\phi}\right)^2\right]} \tag{4.17}$$

and

$$k_l = \frac{\sin(\phi-\beta)}{\sin\phi} \tag{4.18}$$

In these equations α is the angle of the bank to the horizontal, ϕ is the angle of repose of the submerged granular material and β is the angle of the longitudinal slope (i.e. the longitudinal slope S = $\sin\beta$). Values of angles of repose are presented in Section 2.1, Table 2.2.

For flow conditions outside those listed above, the designer should follow the recommendations presented in Sections 2.3.2 and 2.4. Laboratory tests may prove necessary for the determination of block stability under extreme or very specific flow conditions, and/or for block types with substantially different geometries to those listed in Appendix 1.

As for loose blocks, it is important to ensure adequate drainage under the block mats, by means of either granular or fabric filters — see Chapter 5 for guidance on design. Provision of suitable restraint at toes of banks and edges is also of paramount importance.

Suitability

- Research studies (CIRIA, 1987, Escarameia, 1995) have indicated that well-laid concrete block mats can provide adequate erosion protection for current flows with velocities in excess of 4 m/s. This is combined with an ability to withstand wave attack and flow turbulence, which makes them particularly suitable to protect the banks and beds of tidal channels and rivers, and also reaches at some distance downstream of ungated and gated weirs, bridges, etc.
- In the above situations, it is generally advisable to adopt cellular blocks rather than solid units, to promote the release of pressure that may build up behind the revetment. Furthermore, vegetation may naturally establish or seeds can be sown in the openings of the blocks at the upper bank reaches.
- On the other hand, some proprietary solid block mats have textured surfaces that will more easily provide suitable environments for molluscs and plants in tidal areas.
- Being built to cover several square metres with each panel, cabled block mattresses can therefore be ideal for the protection of large areas with good access for lifting cranes; their economic advantage becomes less apparent for schemes of small dimensions.

4.3.3. Composite types

As for other hard revetment systems, concrete blocks are increasingly being used in conjunction with other materials in order to soften their sometimes harsh appearance, to enhance the environmental conditions for flora and fauna development or to fulfil specific needs. Among the latter, is the need to guarantee certain minimum river or channel widths, which can sometimes be best achieved by vertical sheet piling near the toe and upper bank protection by concrete blocks (see Figure 4.14).

The simplest form of composite revetment featuring open-cell concrete blocks consists of cells filled with topsoil and grass seed above mean water level or along the whole of the revetment (in the case of channels with temporary flows). Field trials described in CIRIA (1987) demonstrated that the presence of a well established grass cover can enhance the stability of the block revetment by two means:

Figure 4.14. Example of composite revetment formed by low level sheet piling, fibre rolls and grassed concrete blocks (courtesy of MMG)

- lateral restraint of the soil/roots within the joints of blocks
- grass root anchorage through the base (only possible when granular or woven synthetic filters are used).

When the above two actions are mobilised, it was found that concrete blocks can withstand flow velocities up to 8 m/s (for perimeter contact of more than 75% of block perimeter) and 6 m/s (for perimeter contact between 40 and 75%). These values apply to blocks with the following characteristics: minimum superficial mass of 135 kg/m^2, minimum block mass of 15 kg, minimum block thickness of 85 mm and minimum block dimension in plan not less than three times the block thickness. It is important to note that regular maintenance (albeit involving only simple grass-cutting machinery) will be required in the growing season.

This solution, and any others involving vegetation can only be successful (or economical) in climates that permit vegetation growth without substantial effort. Obviously, the choice of plants also dictates the success of a scheme (native species always being preferred) and therefore careful consideration should be given to the selection of plant species. Comprehensive information on suitable plants for UK waterways is outside the scope of this book but can be found in publications such as Morgan *et al.* (1998), Hemphill and Bramley (1989) and in the relevant references presented in these publications.

Common composite solutions involve the use of low-level piling and planted fibre rolls at mean water level. The upper bank protection is achieved by using fairly standard loose concrete blocks, or concrete blocks below mean water level combined with grass cover on the upper reaches.

4.4. OTHER REVETMENT TYPES

4.4.1. Bitumen-bound materials

Bitumen is a chemically inert and viscous mixture of hydrocarbons that occur naturally or are a by-product of the petrochemical industry. Among its distinguishing properties, is its demonstration of thermo-plastic deformation, which means that the deformation of bitumen depends on the temperature and on the duration of the loading. Standard penetration tests are used to classify the different types of bitumen: for example an 85–100 bitumen has a penetration value between 85 and 100 mm.

The use of bitumen can improve the stability of loose materials used in river and channel revetments. The products available cover a very wide range of functions: from the impermeable lining of canals to the thin coating of sand and stone particles. The viscous nature of the material makes it ideal for binding stones and blocks while retaining most of the flexibility of the revetment. On the other hand, bitumen can also be used in concrete to produce a very dense, impermeable mix. Bituminous revetments have been widely used in the Netherlands as they present an efficient means of reducing the amount of stone required (for the same flow conditions, substantially smaller stone sizes are needed when they are bound with bitumen). In the UK, this type of material has proved less popular, probably due to three main factors:

- the availability of cost-effective alternatives
- the perception that specialised construction plants are required
- the 'unnatural' appearance of bitumen.

With the availability of pre-fabricated asphalt mats, the second factor is no longer very relevant and there is plenty of evidence that vegetation covers can successfully establish through some bituminous revetments. Also, from a water quality point of view, there should be little concern because bitumen is chemically inert. However, it should be borne in mind that bituminous mixes are spread at high temperatures, of the order of 100°C, which can have a temporarily damaging effect on local flora and fauna and produce an oily look to water lying on the surface of the revetment. These aspects should be balanced against the benefit of having a long lasting bank and bed protection system (with a design life of the order of 20 years or more).

Bitumen is used in a number of different types of revetment for rivers and canals, but these types can be divided into two main categories according to their porosity: permeable and impermeable revetments. A summary of the main properties of bitumen-bound materials is shown in Table 4.4 and a more detailed description of each type is presented below.

Permeable types

Like all porous revetments, permeable bituminous types are particularly suitable for environments where waves, tidal flows and other differential loads are

Table 4.4. Bitumen-bound materials — typical properties

Types	Density: kg/m^3	Permeability	Minimum layer thickness
Permeable types			
Open stone asphalt (*in situ*)	1900–2000	e^* = 20–25%	2 to 3 × maximum stone size
Open stone asphalt mats	2000	Same as geotextile used as base layer	2·5 × maximum stone size
Sand asphalt	1500–1800	Same as sand	0·15 m above water 0·70 m below water
Impermeable types			
Dense stone asphalt	Similar to stone	Impermeable	2 to 3 × maximum stone size
Asphalt concrete	2400	Impermeable	2 to 3 × maximum aggregate size
Mastic	2000–2200	Impermeable	0·08 m
Asphalt grouted stone	Dependent on composition of mix	From impermeable to $e^* \approx 30\%$	—

* Voids ratio, e = volume of voids/volume of solids.

combined with current attack. They may also allow the growth of vegetation through the openings of the revetment (if these are sufficiently large) and therefore provide an environmentally acceptable revetment in many instances. The dark appearance of the bitumen tends to soften with time thereby improving the aesthetics of the revetment (see Figure 4.15).

Figure 4.15. Open stone asphalt (OSA) showing naturally occurring vegetation (courtesy of Hesselberg Hydro)

Various degrees of permeability can be achieved depending on the sizes and combination of sizes of the particles coated.

(a) ***Open stone asphalt (OSA)*** is an open structure mixture formed by crushed stone bound by a bituminous mortar (mastic). The usual gradings of stone are 20/40 mm and 16/22 mm, and represent 80% (mass) of the mixture. The remaining 20% is formed by mastic, which is itself a mixture of 60–70% sand, 20% bitumen and 15–20% filler. The coating of the stone is only 1–2 mm thick, thus producing a large voids ratio of the order of 25%.

The construction of an OSA revetment usually involves two stages: the preparation of the mastic and the mixing with the pre-heated stone. Loose open stone asphalt is normally placed above water; for underwater applications prefabricated proprietary OSA mats have been developed with similar characteristics to in-situ OSA (see Figure 4.16). The permeability of the mats is, however, largely dependent on that of the geotextile used as the underlayer.

Design

Guidance on stable revetment thicknesses of OSA is given in Table 4.5, which was adapted from PIANC (1987), and can be used in the initial design stages. OSA revetments usually require an adequate filter layer (or layers) to prevent migration of fines out of the base soil through the open structure of the cover layer. The two most common solutions involve either a geotextile (see Chapter 5) or a sand asphalt granular layer (described later in this section).

Figure 4.16. Placing of open stone asphalt mat underwater (courtesy of Hesselberg Hydro)

Table 4.5. Typical thicknesses of open stone asphalt layers in mm (from PIANC, 1987)

	Bank protection		Bed protection
Situation	In situ	Mats	Mats
Small rivers and channels with restricted navigation	100–150	80–20	100–150
Large rivers, estuaries and navigation channels	150–250	150	150

As a result of project experience, the following rule of thumb was developed by Hesselberg Hydro of the Netherlands, which can be useful in the preliminary stages of design for revetments subjected to wave attack:

$$t = CH_s \tag{4.19}$$

where
t is the OSA thickness
H_s is the significant wave height (see Section 2.3.2)
C is a coefficient dependent on the underlying material

C = 0·17 on geotextile
C = 0·10 on sand asphalt.

It can be seen that in river situations, where wave heights will seldom exceed 0·5 to 1 m, this equation generally gives smaller thickness values than Table 4.5. Although both guidelines are considered to be of interest, particularly the latter which takes into account the nature of the underlying soil material, a conservative approach is generally advisable. Minimum recommended thicknesses of about 2·5 times the maximum stone size should always be achieved.

A more accurate estimation of the layer thickness can be obtained by considering that an asphalt layer on a soil foundation subjected to wave attack deforms like a plate on an elastic base. The thickness of the revetment must be such that it resists the bending moments caused by the waves. For a design equation based on this conceptual model refer to PIANC (1987).

Suitability

- Research in the Netherlands has shown that OSA can withstand very high flow velocities, of the order of 7 m/s, without suffering significant damage.
- It can also provide adequate stability to river beds and banks under severe wave attack.
- The above characteristics make it particularly suitable where extensive protection is required in large navigable rivers and canals. Natural vegetation has been found to establish on OSA and provide an aesthetically pleasing appearance to the revetment.

(b) ***Sand asphalt*** consists of sand coated with 3–5% of bitumen. Its permeability is very much that of the sand of which it is composed, but some degree of cohesion is provided by the bitumen. As well as being used as a revetment in its own right, sand asphalt is in many instances applied as a temporary revetment and in filter layers to help to stabilise slopes.

Bulk asphalt is a type of sand asphalt with a slightly higher proportion of bitumen (6% by weight) which has been successfully used on upper banks of rivers like the Mississippi in the USA.

Design

Sand asphalt should be placed at slopes of the order of $1V{:}3H$. It can be placed both above water (with a recommended minimum thickness of 0·15 m) and below water (minimum thickness of 0·70 m).

Suitability

- Sand asphalt is used both as a revetment and as a granular filter layer, in many cases underlying open stone asphalt.
- A suitable application as a revetment consists of the protection of upper banks of rivers with currents below 2 m/s.
- It may require fairly frequent maintenance as sand asphalt layers can be worn by abrasion, at rates of up to 3 mm/year. They can also silt up with time and therefore increase the risk of failure being caused by differential hydrostatic pressures.

Impermeable types

Some of the materials described in this section, such as mastic and asphalt grout, are not always used as revetments on their own, but are important constituents of revetment layers. In some cases, impermeable bituminous revetments have to be able to resist forces caused by turbulence fluctuations, waves, variable water levels, etc. In situations where these forces are relatively small, impermeable bitumen-bound materials can provide the necessary constraint to loose cover materials and at the same time produce an even, integral surface. Weepholes are sometimes introduced to minimise the build-up of pressures underneath or behind the revetment. Compared to cement grout, bitumen offers considerably higher flexibility to the revetment, thus allowing some differential settlement before damage occurs. A description of the major types used in river engineering is presented next.

(c) ***Dense stone asphalt*** is a mixture of 50–70% stone and mastic (which is a mixture of 60–70% sand, 20% bitumen and 10–20% filler). Usual gradings of stone are 20/40 mm and 16/22 mm. The hot mixture can be poured or hosed both above and below water.

Design

The minimum thickness recommended is 2 to 3 times the maximum stone size, being the design thickness determined by stability analysis of impermeable revetments (see Section 4.4.4) or by specialist technical advice. It is generally recommended that an adequate underlayer (geotextile or granular filter) be provided, which has the principal function of draining the zone parallel to the slope of the revetment.

Suitability

- Dense stone asphalt provides adequate protection for moderate to heavy current attack of the order of 2–5 m/s.
- Its impermeability (which can be controlled by the introduction of weepholes for the release of excessive pressures underneath the revetment) renders it mainly suitable for bank revetments above the high water level. It can also be placed under water for bed protection in cases where small uplift pressures are expected.

(d) ***Asphalt concrete*** comprises 50–60% crushed stone or gravel, 7–8% sand and filler and 7% bitumen. It is used as an impermeable revetment above the high water level; its application below water level should be restricted to non-tidal and non-navigable rivers and channels.

Design

Little information is available on the design of asphalt concrete. Although having greater flexibility than ordinary concrete revetment layers, its design thickness can be determined using general principles of stability for impermeable revetments (see Section 4.4.4).

Suitability

- Asphalt concrete is suitable for sites subjected to high flow velocities (above 2·5 m/s) but small differential pressure (since the revetment is impermeable).
- It is preferably used above high water level.

(e) ***Mastic***, which is formed by 60–70% sand, 20% bitumen and 15–20% filler, is a very flexible product that can be poured when hot. It provides an impermeable layer for river banks and beds as well as being very suitable for toe protection. When combined with stone (as in open and dense stone asphalts) the quantity of mastic is reduced and its flowing properties can be more effectively controlled.

(f) ***Asphalt grout*** is formed by a mixture of mastic (see description above) and gravel or crushed stone up to 50–60 mm in size. Being more flexible than cement grout, it is used to fill the interstices of riprap, hand pitched stone or other revetment systems while retaining some of the flexibility of the loose material. Depending on the proportion of the holes that are filled with grout, the resulting revetments may have various degrees of permeability: these range from totally impermeable (full grouting) to a partial permeability (surface and pattern grouting).

In order to reduce the risk of grout running down the slope during application, it is recommended to limit the bank slopes to $1V{:}2H$.

4.4.2. Flexible forms

(a) ***Filled sacks.*** Flexible forms such as hessian sacks filled with concrete have long been used for river bank protection in the UK (see Figure 4.17). In most situations they are piled up to form near-vertical walls with practically no inter-unit bonding, in a way that resembles traditional dry-stone walls. Various types of filling that are used include concrete, mortar and sand. Sand filling is particularly suited for emergency or temporary works (e.g. rising of flood defence banks) where rapid and inexpensive filling of the sacks is the prime concern.

The sacks are usually laid stretcher-bond and can also be pinned down with steel bars to form a lined bank at a gentle to medium slope. Common bag sizes are 600 mm in length, 300 mm in width and 150 mm in thickness when filled.

Filled synthetic sacks (which are more commonly available today than hessian sacks) offer a cheap solution that allows manual placement; however, they have serious limitations from the environmental and aesthetic points of view. Although some organic growth has been found to establish over the years, this is generally limited to thin slime covers. Visually, they tend to rate badly when compared to other options. However, this is a subjective issue since it has been found that some residents in areas where sacks have been in place for several decades have grown accustomed to their appearance. In these cases, when repairs are needed, it has been decided to replace the damaged areas with similar products. Concrete filled bags can also be used as cappings for steel sheet piling to provide a transition to the above-water bank, where bioengineering revetments can be adopted.

Figure 4.17. Concrete-filled sacks installed approximately 60 years ago (note collapse possibly due to boat wash)

Design
Being essentially a traditional revetment used for temporary works, the design of filled sacks has been based more on experience and common sense than on stability considerations. As mentioned above, two different situations can arise which require different design approaches.

- Situation 1: the sacks are laid on a slope.
- Situation 2: the sacks are stacked to form a wall.

In Situation 1, in order to provide protection to the toe of the bank, it is advisable that the lowest layer or layers of concrete-filled bags be placed in a trench dug along the toe. The use of bagwork on slopes is now generally restricted to the localised repair of collapsed banks and design procedures are limited. Steel bars are sometimes spiked through the bags for added stability.

Situation 2 should be regarded as a gravity wall case. The stability of the wall should then, in principle, be assessed for all the relevant modes of failure (see, for example, Hemphill and Bramley, 1989). However, there may be difficulties in establishing the value of the frictional resistance between bag layers and therefore in carrying out a full design. A simplified approach that has been followed by some designers consists of first calculating the main destabilising forces and moments; these are then assessed, and, if considered severe, steel reinforcement bars are used to spike through the bags. The number and spacing of the bars is determined from past experience. In order to limit the uncertainty of the design, it is recommended that the number of bag layers does not exceed 3 to 4. For an average bag thickness of 150 mm, this corresponds to a limiting wall height of the order of 0·5 m, although this limit is often exceeded in practice.

Suitability
- Filled sacks or bags are primarily materials for use in the repair of existing walls or for emergency situations, particularly to increase the height of flood defences. However, they have also been used historically to protect banks of small streams and rivers from erosion.

(b) ***Flexible form mattresses.*** Flexible form mattresses basically consist of woven synthetic fabric cases filled with concrete (or other materials such as sand and mortar) where the fabric acts as a lost shutter (see Figure 4.18). The fill is pumped into the mattress, which provides the restraint and protection against erosion that is needed until the concrete has set. Being porous, the mattress fabric allows the slow release of excess water from the concrete mixture, thus increasing its strength and density.

High strength woven textiles such as polyester, nylon and polypropylene are generally chosen for the mattresses, which can be made in a wide range of shapes. This latter feature, as well as the flexibility of the product before filling, makes it suitable not only as a revetment for river beds and banks but also for repairs to hydraulic structures, particularly for underwater situations. Protection of river and channel beds in fast flows or highly turbulent environments can be successfully achieved with flexible form mattresses with internal reinforcement and adequate bolting to nearby fixed structures. Some proprietary systems offer

Figure 4.18. Example of flexible form mattress in a tidal river (courtesy of Proserve Limited)

the possibility of designing the base fabric layers so that they function as geotextiles and thus prevent the loss of solids from the underlying ground.

Although cement mixes such as micro-concrete (a fluid concrete with inerts $< 1{\cdot}5$ mm) and mortar are the most common fills, materials such as slags, sand and bitumen can also be used. Prototype tests carried out in the Netherlands with sand-filled mattresses have highlighted the importance of a sufficient degree of compaction of the sand inside the mattresses. The necessary density of fill may be difficult to ensure if the filling equipment does not function properly. Corrective filling may then be required. Once filled, the revetment surface usually has an undulating appearance with small 'pockets'. These can encourage the deposition of silts in tidal regions. Some limited organic growth has been found to establish on this type of revetment.

In order to relieve hydrostatic and differential pressures in what is essentially an impermeable revetment, some concrete-filled systems allow for the introduction of apertures. This can be a requirement in tidal reaches, in navigable channels and other situations where waves need to be taken into account in the design. The flexibility of the revetment can be increased in some proprietary systems by good articulation at the joints of individual compartments in the mattresses and also at their perimeters. Special zips, straps, bolts and ties are used to secure mattresses to adjoining ones.

Design

The major manufacturers of flexible form mattresses can usually offer comprehensive technical information on their products and provide full design services. Due to the specialist nature of this type of revetment, this is probably the best approach to take in terms of design. However, it is important to note the following points. When, as is most common, the mattresses are concrete filled, they form a revetment which differs little from an in-situ concrete lining. Therefore, they are impermeable (unless pressure relief holes are purposely introduced). Where significant water level variations are expected (e.g. tidal flows, seasonal variations in tropical countries, waves) the dissipation of uplift pressures from behind the revetment must become a priority (see Sections 2.3.2 and 2.3.3).

In situations where flexibility and permeability are of paramount importance, sand-filled mattresses may be the preferred option, particularly in sites where rock is unavailable. Local labour can generally be used to lay the mattresses and carry out the filling. This type of mattress with a design weight of 200 kg/m^2 has been adopted for bank protection under currents of the order of 2·5 m/s.

Although the mattresses are designed to be sufficiently robust not to be damaged by pins driven through them, in navigable waterways it is advisable to build mooring jetties at regular intervals. The top layer of mattress generally deteriorates with ultra-violet light and can become lost after 20–30 years in UK conditions; in equatorial climates the process is estimated to accelerate to less than 10 years for above-water applications.

Suitability

- Flexible form mattresses are suitable for the protection of long reaches (above and under water) and for repairs in the vicinity of hydraulic structures (e.g. culverts, weirs and ship docking areas).
- When filled with concrete they can help to prevent erosion of river and canal bed and banks under heavy current attack (velocities above 2·5 m/s).
- When filled with sand or other granular materials, they have been found to withstand velocities of up to 2·5 m/s, but attention should be given to the possibility of damage by vandalism to some proprietary systems. Other problems can occur if there is difficulty in ensuring sufficient compaction of the fill.

4.4.3. Soil reinforcement systems/geomats

This category comprises two types of system that are used to increase the stability of banks: soil reinforcement or confinement systems (commonly having a honeycomb structure) and three-dimensional geotextiles (geomats). Two-dimensional geotextiles are described in detail in Chapter 5 and can be broadly distinguished from geomats by their different functions in a revetment system. While geotextiles have essentially a filtering function, geomats can also retain soil or other materials to form a revetment on their own. Examples of soil reinforcement systems and geomats are shown in Figure 4.19 for illustration purposes. A great number of proprietary systems are available in the UK and data sheets for some of the most common products are presented in Appendix 1.

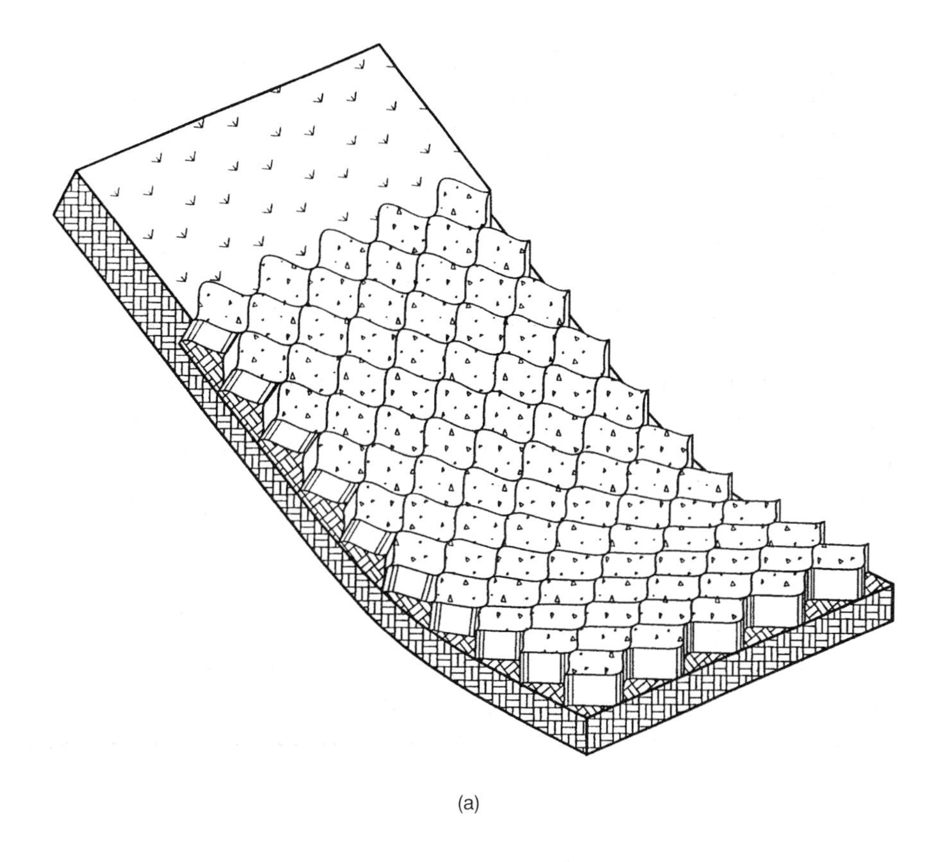

(a)

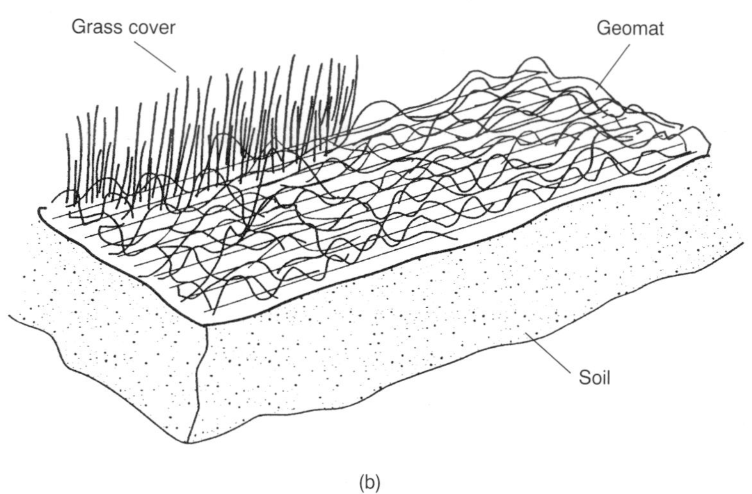

(b)

Figure 4.19. Examples of soil reinforcement system and of geomat: (a) soil reinforcement system; (b) geomat with grass cover

(a) *Soil reinforcement systems.*

Soil reinforcement systems (also known as cellular confinement systems) are based on the principle that stability of soils can be enhanced by increasing their shear strength and stiffness. The cell walls provide strength and frictional interaction with the fill material, while adjacent cells give passive resistance. This improvement in strength allows a reduction in the depth of the revetment and/or of its foundation. Being expandable, these systems do not involve high transportation or handling costs but it is important to have good joining between cells. In most cases, the cells are made from strips of plastic material (although geotextiles and fibres have also been used) and are of varying heights and plan sizes; the strips can either be sewn together or welded.

Some proprietary systems offer a choice of colours to match the surrounding area and a range of wall finishes (from smooth to textured and perforated) to promote adequate bonding with the fill. This can consist of earth (and particularly seeded topsoil), stone or concrete, depending on the strength of the flow.

In some proprietary systems additional strength is mobilised by introducing tendons in aggregate and concrete-filled systems. This can be a requirement in steep slopes, in cases where the underlying soil is too hard to drive anchor stakes, or where impermeable membranes are used that should not be punched. Figure 4.20 illustrates two types of application of cellular confinement systems: as a channel lining in Figure 4.20(a) and as a combination of revetment and retaining wall in Figure 4.20(b).

Design

The design of soil reinforcement revetments usually involves the choice of cell sizes and depth, the nature of infill(s), the type of filter and the method of anchorage. It can generally be said that cell sizes and depths depend directly on the maximum size of the infill particles. It is normally accepted for perennial rivers and channels that soil infill with grass cover should be restricted to upper bank reaches; below water level the options include concrete and stone. However, in intermittent flow channels the use of grass can be extended to the whole of the channel perimeter.

The choice of infill is mainly dictated by the characteristics of the flow environment but can also be heavily dependent on the availability of suitable materials. Although graded stone has been found to provide lower stability than uniform stone with the same D_{50} size, it may be chosen for practical reasons. Some proprietary systems can provide limiting values of flow velocity to adopt for design that are based on research studies.

As a very general rule of thumb, soil infill with well established grass cover may withstand velocities in the order of 4 m/s for short periods of time (less than 2 hours, typically). The stability of concrete-filled systems is heavily dependent on the cell depth, i.e. on the thickness and weight of the revetment. For example, 150 mm deep cells may successfully cope with peak flow velocities in the order of 6–7 m/s.

The placement of a geotextile underneath the cellular structure is recommended by most manufacturers. Reasons for this include the need for separation between the fill material and fine-grained soil, and prevention of upward movement of fines into the fill due to variation in porewater pressure behind the revetment.

Several anchorage methods can be used, such as steel pins at regular intervals (in steep slopes as frequently as 1 pin/m^2 may be required), stake anchors, dead

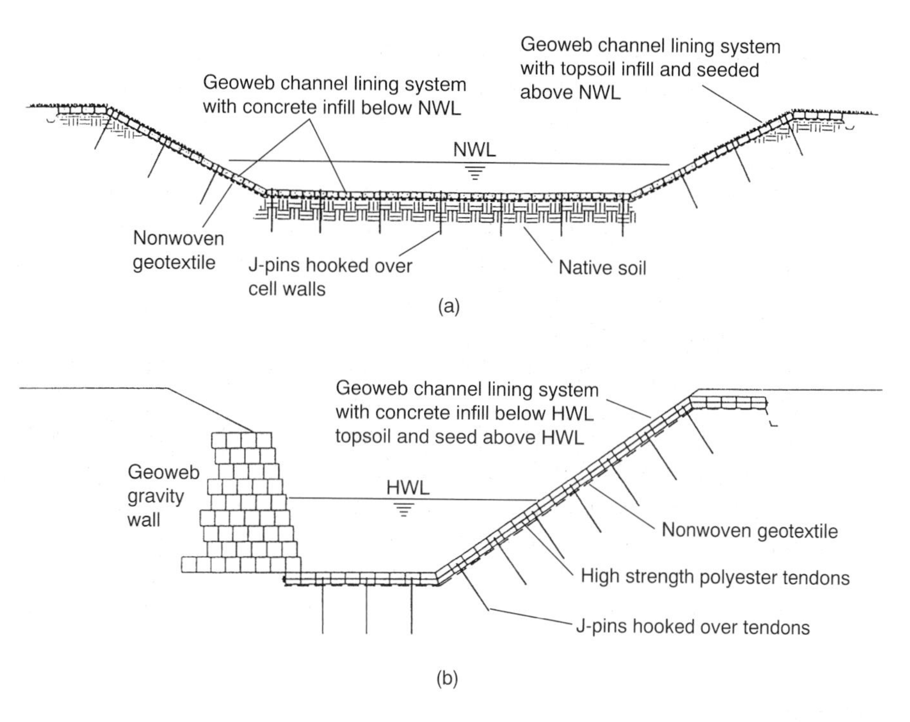

Figure 4.20. Examples of soil reinforcement systems (adapted from Geoweb commercial literature); (a) channel lining; (b) retaining wall combined with slope and bed revetment

man anchors (i.e. buried steel or timber beams) and trenches (usually to fix the systems at the top of the bank). In concrete-filled systems adequate provision of pressure relief weepholes should also be considered.

Suitability

- Soil-filled cellular systems are suitable for protection of upper banks.
- Concrete and stone fills can successfully be used below water level. Vertical or near-vertical walls can also be built by stacking several layers of soil reinforcement systems.

(b) ***Geomats.*** Geomats are described here as three-dimensional geotextile membranes, which have sufficient thickness to retain soil or other fills, and that may allow the growth of vegetation through the openings. In the early stages, the function of geomats is primarily to assist in the establishment of a suitable vegetative cover. Later, they provide reinforcement to the plant roots and increase their natural resistance to erosion (see Figures 4.19(b) and 4.21). Good contact with the underlying soil is important if these objectives are to be achieved. However, since this is not always easy to check on site, it is good practice to cover the geomat with a 30–50 mm thick layer of topsoil. The installation should be carried out at a time of the year when vegetation is likely to be established before flood risk is high.

Usually, geomats are formed by randomly entangled filaments of plastic material and are only a few millimetres or centimetres thick. Some proprietary systems include a variety of fills in their range of prepared geomats, such as pre-

Figure 4.21. Example of geomat placed on a bank (courtesy of Maccaferri Ltd)

cultivated grass for instant protection against erosion and bitumen-bound gravel for heavier current attack. On the other hand, some products are fairly thin but are designed to increase erosion resistance by allowing root growth through the openings; they can be used in conjunction with cellular soil reinforcement systems, for example.

Design

The capacity of geomats to resist flow attack relies greatly on good preparation of the underlayer and on construction provisions to avoid detachment from the base soil. If this is not achieved, collapse of the revetment may occur due to the relatively low weight of these products. It is therefore also recommended to start laying mats (which are usually supplied in rolls) from a downstream position, to avoid upstream-facing overlaps. Fixing of the mats is generally carried out by anchoring with pins or burying in trenches at the top and bottom of banks. Typical trench sizes are 450 mm wide by 250 mm deep.

Several different types of geomat with a wide range of erosion resistance characteristics are available to the designer. Generalisations are therefore difficult to make, especially since most of the proprietary products have not been tested for their hydraulic resistance in a systematic way. Some basic guidelines are, however, given in Table 4.6 to assist the designer, based on research studies (e.g. CIRIA, 1987) and on manufacturers' information.

Table 4.6. Hydraulic resistance of geomats

Type of geomat fill	Flow velocity: m/s
Geomat without vegetation cover or bitumen-bound gravel	$\leq 1{\cdot}5$
Geomat with bitumen-bound gravel fill	$\leq 2{\cdot}5$
Geomat with well-established grass cover	$\leq 3{\cdot}5$
Geomat with bitumen-bound gravel and grass cover	≤ 5 (or higher in some cases)

Suitability

- Geomats are primarily suitable for protection of the upper banks against erosion, above mean water level.
- Some proprietary systems can also provide adequate protection in situations of moderate to high current attack, particularly when filled with materials such as bitumen-bound gravel.

4.4.4. Concrete

As one of the most effective types of construction material, concrete has also been extensively used in river bank and bed protection. It offers strength, durability, chemical inertia in most environments and can be made to follow irregular contours or take any geometric shape. Concrete is therefore found in vertical walls of canals and channelised rivers in urban areas, in the lining of watercourses, in toe protection, at edges of revetments or as a grouting material for revetments such as stone, flexible form mattresses, gabions, etc. Figure 4.22 shows an example of an urban watercourse lined with concrete; although

Figure 4.22. Concrete-lined channel

probably fulfilling its hydraulic and geotechnical functions adequately, its unattractiveness is quite apparent, even though some low vegetation cover has managed to establish itself on the bank on the left-hand side of the photograph.

On the other hand, concrete revetments (and by this solid in-situ lining and pre-cast slabs are meant) involve high initial investment costs, are rigid and impermeable and, more importantly, they do not rate highly in environmental terms. For these reasons their popularity has declined dramatically in the UK in the past 20 or 30 years: concrete has been replaced by methods and materials more able to sustain animal and plant species as well as being aesthetically superior. In spite of the above, it is important to note that environmental considerations do not have the same weight all over the world — revetment choice can be dictated primarily by economic or hydraulic arguments, for example. The next two sub-sections deal with in-situ concrete lining and pre-cast slabs; concrete gravity walls are not covered in this book (see, for example, Hemphill and Bramley, 1989, for information). Brief information on the use of concrete as a grouting material can be found in Section 4.1.4 (and also 4.2), whereas block revetments made of concrete are described in detail in Section 4.3.

(a) ***In-situ lining.*** In-situ concrete lining can assume two main different forms, solid or cellular concrete slabs, the choice depending on local soil and flow conditions as well as on particular requirements. In both cases, steel bar reinforcement is generally adopted to prevent excessive cracking.

Cellular slabs can be proprietary (see Data Sheets in Appendix 1) or built on site with appropriate formwork. The cell formers are present during construction and are burnt out afterwards once the concrete has set, leaving spaces that can be filled with topsoil and later provide habitat for vegetation. The negative visual impact normally associated with concrete lining can thus be softened.

Design

The following important issues need to be considered in the design of in-situ solid concrete slabs:

- due to the impermeable nature of this revetment, it is essential to make provision for a sufficient number of weepholes or other means of releasing groundwater pressure, and
- in order to allow some differential foundation settlement and accommodate temperature changes, the slabs need to be provided with joints at regular intervals.

These requirements assume particular relevance when lining river or canal beds with solid slabs downstream of hydraulic structures. In this situation, the high turbulence levels can generate extremely strong up-lift forces and ultimately lead to collapse.

Minimum slab thicknesses are usually around 100–150 mm and bank revetment without top formwork is limited to slopes of $1V{:}1{\cdot}5H$ or flatter for construction reasons. Further guidance on concrete design (including reinforcement) should be sought from specialist sources as it is outside the scope of this book.

Suitability

- With adequate design, which should include provision for pressure relief and small differential movement of the foundation, in-situ concrete linings can be used in heavy current attack.
- It is a durable system, suitable for both bank and bed protection, particularly if some allowance is made to promote vegetation growth above the mean water level (e.g. by using cellular slab revetments or by combining solid slabs with soft protection in the upper bank, if possible).

(b) ***Pre-cast slabs***. The distinction between a concrete block and a pre-cast slab is not universally established and therefore a general definition does not exist. The difference lies, basically, in the ratio of the thickness to the surface area of the unit. Being slenderer than blocks, slabs have smaller thickness/area ratios, typically smaller than 0·075 per metre, whereas blocks have ratios typically greater than 0·6 per metre. Also, pre-cast slabs are regular in shape (rectangular) and, in most cases, solid (see Figure 4.23); however, in some countries, such as Poland, cellular pre-cast slabs are commonly used in river bank protection. Blocks usually feature irregularities at the perimeter to promote interlocking and are fabricated with a cellular structure as well as solid.

Design

Due to their relatively small weight per unit area, pre-cast slabs require careful design to ensure stability under current flow. Even in straight channels with no appreciable wave generation there can be scope for the development of excessive uplift forces leading to the collapse of the channel lining. This scenario may arise when the inward and outward movement of tides, or the release of flow from pumping stations creates a difference between the pressures on the outer and inner faces of the lining. Pressure relief holes can be introduced to minimise this

Figure 4.23. Example of channel protection using pre-cast concrete slabs (note sand leaching from underneath)

problem but it may be wiser to adopt a different type of revetment in these situations. The design should also ensure that leaching of fines, underlying the slabs, cannot take place through the slab joints, as this could lead to differential movements and failure.

Slab thickness is determined by balancing the stabilising forces (usually weight, frictional resistance of the underlayer, lateral constraint) and the destabilising forces (current attack, uplift). As a general recommendation, it is advisable to avoid thicknesses smaller than about 80 mm. Care must be taken to provide good toe support at the base of a bank.

Suitability

- Provided that they are properly designed, pre-cast slabs can be used for the lining of channels and rivers in much the same way as loose solid concrete blocks with no interlocking. This means that they are suitable for areas where waves and turbulence levels are low.
- This revetment system has the disadvantage that, once local failure has been initiated, progressive failure can rapidly follow.
- Although the surface of the slabs can be treated to improve their appearance and encourage growth of slime, moss and other small species, this type of revetment will seldom be very environmentally friendly in temperate countries. The application of pre-cast slabs in arid or semi-arid regions is probably more acceptable.

4.4.5. *Piling*

Piling is one of the possible methods used in rivers to build vertical walls for protection against erosion. Mention has been made in previous sections of systems that can be used as revetments as well as retaining walls (e.g. block stone in Section 4.1.2, box gabions in Section 4.2.1 and filled sacks in Section 4.4.2 (a)). However, in rivers with limited available space for bank protection, the most efficient solutions often involve the use of piling. Since vertical walls are of subsidiary interest to this book, the information given here is intended only as a basic introduction to piling. The subject is also a specialist one, and the river engineer is advised to seek expert advice on design. A good summary of the use of piling in rivers and canals can be found in Hemphill and Bramley (1989); for the design of sheet piling, publications such as EAU (1992), British Steel (1997) and BS 8002 (1994) *Code of practice for earth retaining structures* provide comprehensive technical information. A recent publication by the UK Institution of Civil Engineers (1996) gives guidance on specification. Software is also commercially available for the design of retaining walls, including sheet pilling.

Sheet piling is used both in temporary river works (such as cofferdams for the construction or repair of weirs, gates, groynes, and the local protection of banks during major river engineering), and in permanent schemes (in most cases for the protection of river banks in urban areas or other restricted areas, close to hydraulic structures and in navigable rivers or canals)—see Figure 4.24. Reasons for choosing piling include not only limited space but also the need to protect banks against severe flow conditions. They are also suitable in situations of

(a)

(b)

Figure 4.24. Piling: (a) example of piled wall; (b) protection of river banks using sheet piling near a gate

restricted land access, since piles can be driven from barges. It is important to note, however, that sheet piled walls may require tensile support in the form of dead man or ground anchors, which need to be located outside any potential failure surfaces. Figure 4.25 illustrates the two main types of sheet piled wall, cantilever and anchored, and also includes features that can be used to enhance their environmental characteristics.

Steel is the material most frequently used in sheet piling: it can be driven without suffering appreciable deformation, it is a very durable material even in saline waters and contaminated soils, and sufficient strength is achieved with sections only a few centimetres thick. In the UK, lengths are available up to 25 m, with thicknesses of up to 0·0254 m. In the majority of cases it can be used unprotected, although it is sometimes painted for aesthetic reasons. In the situations covered in this manual, the rates of corrosion are greater where the wall is backed by soil and is subjected to splash on the front face (0·09 mm/year). For most cases, a rate of corrosion of 0·05 mm/year can be assumed in design (British Steel, 1997). Serviceable lives of 60 years or more are common for unprotected steel piling. Another form of corrosion, bacterial corrosion, has recently been identified as a potential danger to many sheet piling structures in saline environments. This form of corrosion, known as accelerated low water corrosion (ALWC), can apparently erode piles at a rate of up to 1 mm/year in both temperate and tropical climates and, if undetected, can therefore have serious economic consequences.

Reinforced concrete sheet piling can be an alternative to steel where the piles can be driven into the soil without risk of damage and where the wall does not need to be absolutely watertight. Thicknesses of piles range between 0·14 m and 0·40 m and widths are of the order of 0·5 m. Lengths can be made up to 15 m (or 20 m, if required), (EAU, 1992).

Timber sheet piling used in permanent structures should be designed so that the pile tops are constantly wet to avoid decay. Common thicknesses range between 0·06 m and 0·30 m and normal widths are of the order of 0·25 m. A useful rule of thumb for long piles in unobstructed soil states that the wall thickness in centimetres should be twice the pile lengths in metres (EAU, 1992).

Plastic sheet piling is another type that can be used to form relatively low walls (see Figure 4.24a) and is becoming more popular in UK rivers. Like other forms of piling, it may require anchoring to achieve the necessary stability and avoid deformation along the top; due to the limited strength of plastic when compared with steel or concrete, care should be taken not to damage the piling head during construction. In order to improve the visual appearance of piled walls, they are in many cases designed as low walls protecting the banks up to the average water level. Above this level, revetments such as grassed concrete blocks, planted fibre rolls or riprap can be used, as well as bioengineering measures.

4.5. SUMMARY

This section presents two tables summarising some of the main properties of the various revetments described in the preceding sections of this chapter. The first of these tables, Table 4.7, provides qualitative guidance on general characteristics such as flexibility, permeability, need for maintenance and environmental quality. Brief information is also given as to the main areas of application of each type of revetment.

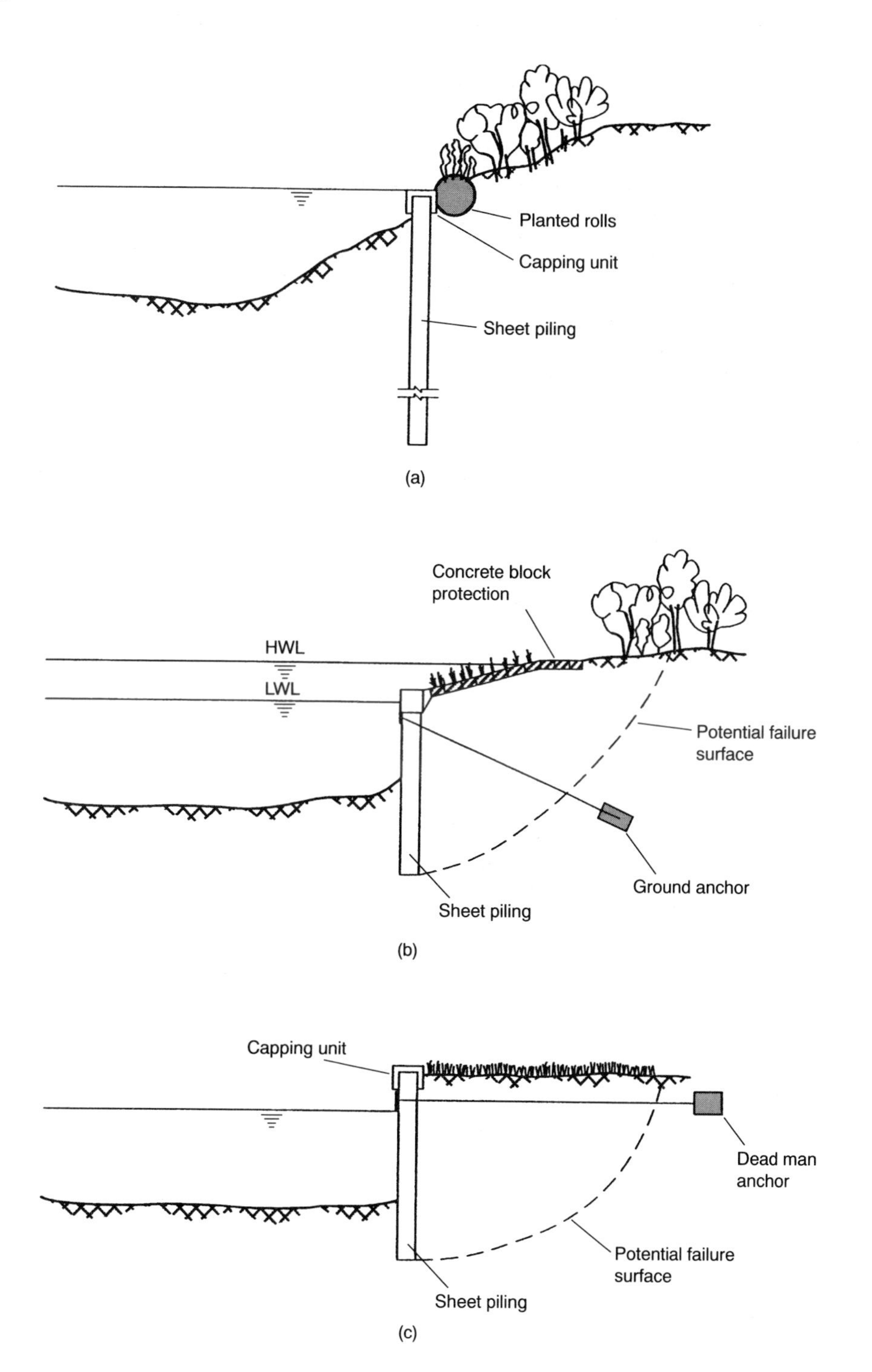

Figure 4.25. Examples of piled walls: (a) cantilever wall; (b) anchored wall (ground anchor); (c) anchored wall (dead man anchor)

Table 4.7. Properties of common types of revetment

Revetment	Flexibility	Permeability	Maintenance requirements	Environmental impact		Main usage
				Visual	Ecological	
Rock						
Riprap	High	High	Low/medium	Good/acceptable	Good/acceptable	Bank and bed protection up to very heavy current and wave attack; also used at transitions and sites with high turbulence levels
Block stone	Medium	Medium	Low	Good/acceptable	Good/acceptable	Upland areas; heavy current attack
Hand pitched stone	Low	Low	Medium	Good/acceptable	Acceptable/poor	Short bank reaches; repairs of existing revetments
Grouted stone (cement)	Low	Low	Low/medium	Acceptable/poor	Poor	Usually associated with hand pitched stone at transitions and other areas of heavier flow attack
Concrete						
In-situ lining						
(a) solid	Low	Low	Low	Poor	Poor	Bed and bank protection in heavy current attack, particularly in urban areas
(b) cellular	Medium	Medium	Low	Acceptable	Acceptable/poor	
Pre-cast slabs	Medium	Low	Low	Poor	Poor	Lining of channels in low turbulence and negligible wave environments
Gabions						
Box gabions						
(a) rock fill	Medium	High	Low/medium	Acceptable/poor	Acceptable	Bank protection downstream of hydraulic structures; retaining walls
(b) soil fill	Medium	Medium	Low	Acceptable	Acceptable	Bank protection; retaining walls
Gabion mattresses	High	High	Low/medium	Acceptable	Acceptable	Bed and bank protection of large areas; protection at hydraulic structures

Table 4.7. (continued) Properties of common types of revetment

Revetment	Flexibility	Permeability	Maintenance requirements	Environmental impact		Main usage
				Visual	Ecological	
Sacks/rolls						
(a) Rock rolls	High/medium	High	Low/medium	Acceptable	Good/Acceptable	Filling of scour holes; toe protection
(b) Bio rolls	High/medium	Medium	Medium/high	Good	Good	Bank protection in low to moderate currents
Concrete blocks						
Loose and/or interlocking	Medium	Medium/ low (solid)	Low/ medium (grassed)	Acceptable/poor	Acceptable/poor	Fairly small areas requiring manual placing; difficult contours
Linked	Medium	Medium/ low (solid)	Low/ medium (grassed)	Acceptable/poor	Acceptable/poor	Above and below water construction
Bitumen-bound materials						
Open stone asphalt	High	High	Low	Good/acceptable	Good/acceptable	Protection of long reaches of rivers (including navigable) above and below water
Sand asphalt	High	High	Medium	Good/acceptable	Good/acceptable	Revetment of upper banks and as filter layer
Dense stone asphalt	Medium	Low	Low	Acceptable/poor	Poor	Mainly above high water level; below water in non-tidal, non-navigable rivers
Asphalt concrete	Medium	Low	Low	Acceptable/poor	Poor	
Mastic	High	Low	Low	Poor	Poor	Mainly used in conjunction with stone
Asphalt grouted stone	High	Medium/low	Low	Acceptable/poor	Acceptable/poor	Used in transitions to improve stability

Table 4.7. (continued) Properties of common types of revetment

Revetment	Flexibility	Permeability	Maintenance requirements	Environmental impact		Main usage
				Visual	Ecological	
Flexible forms						
Filled sacks	Medium	Medium/low	Low/medium	Acceptable/poor	Acceptable/poor	Repair of existing bagwork; near-vertical walls in non-navigable channels
Flexible form mattresses				Acceptable/poor	Acceptable/poor	Particularly suitable for protection of long reaches above and under water
—sand fill	High	Medium	Medium			
—concrete						
(a) before setting	Medium	Medium/low	—			
(b) after setting	Low	Low	Low/medium			
Soil reinforcement systems/geomats						
Cellular systems						
—concrete fill	Medium/low	Low	Medium	Poor	Poor	Used as revetments and retaining walls; below water applications
—stone or sand fill	High	High	Medium	Good/acceptable	Good/acceptable	Upper banks (sandfill) but also on banks below water (stonefill)
—soil/grass	High	High	Medium/high	Good	Good/acceptable	Upper banks
Geomats	High	High	Medium/high	Good	Acceptable	Upper reaches of banks; can be used below water level with appropriate fill material; steep banks

The second table, Table 4.8, details in a more quantitative way, the range of applicability of the revetment types covered in this chapter (plus bioengineered systems, for reference purposes). This table should be used in conjunction with

Table 4.8. Indicative guide to suitability of revetments in normal to medium turbulence conditions (legend to table overleaf)

Mean flow velocity: m/s	Type of revetment	Resistance to wave attack wave height: m
≤1	Bioengineering	Likely to be ≤0·15
	FS	Likely to be ≤0·15
	R20	≤0·15
≤1·5	GP	Likely to be ≤0·15
	SR1	Likely to be ≤0·15
	LB<160	Likely to be ≤0·15
	PC	Likely to be ≤0·15
≤2	SA	Likely to be ≤0·50
	SR2	Likely to be ≤0·50
	LB>160	Likely to be ≤0·50
	HPS	Likely to be ≤0·50
	R100	≤0·50
≤2·5	SG	Likely to be ≤0·50
	GBB	Likely to be ≤0·50
	FFMS	≤0·50
	R150	≤0·75
	R200	≤0·75
≤3	R250, R300	≤1·0
	R400	≤1·5†
≤3·5	GG	Likely to be ≤1·0
	SR3	Likely to be ≤1·0
	GM1	≤1·5†
	R500, R600	≤1·5†
≤4·0	SRG‡	Likely to be ≤1·0
	LBG	Likely to be ≤1·0
	GM2	≤1·5†
≤4·5	CB<250	< 1·5†
≤5·0	GBG	Likely to be ≤1·0
	DSA	Likely to be ≤1·0
< 5·0*	BG, SRC, P, IC, FFMC, GS	Likely to be ≤1·0
	CB>250	≤1·5†
	OSA, GM3	≤1·5†

* Within the applicability of this book (i.e. < 7 m/s).
† Beyond the applicability of this book (i.e. < 1 m).
‡ For periods of time less than 2 hours.
Note: the values given in the above table are *indicative only* and therefore do not exclude the need for detailed calculations. Also, for many of the proprietary products, the values given in this table were obtained from tests commissioned by the manufacturers and tend to relate to the 'top of the range' products available in the market for each category of revetment.

Table 4.7 and be interpreted as an approximate guide for selection in the earlier stages of design. Revetments are grouped according to the mean cross-sectional flow velocities that they can withstand before damage occurs and are related to the maximum acceptable wave height. When preparing the table, some judgements had to be made concerning the stability of particular types of revetment because of incomplete information on certain aspects, such as resistance to wave attack.

Legend:

BG	Box gabions
BS	Block stone
CB	Cabled concrete blocks CB < 250—cabled blocks of weight less than 250 kg/m^2 CB > 250—cabled blocks of weight greater than 250 kg/m^2
DSA	Dense stone asphalt
FFM	Flexible form mattresses FFMC—concrete filled FFMS—sand filled
FS	Filled sacks
G	Geomats GP—plain GBB—with bitumen-bound gravel fill GBG—with bitumen-bound gravel and good grass cover GG—with good grass cover
GM	Gabion mattresses GM1—0·15 to 0·17 m thickness GM2—0·23 to 0·30 m thickness GM3—0·50 m thickness
GS	Grouted stone
HPS	Hand pitched stone
IC	In-situ concrete
LB	Loose concrete blocks LB < 160—loose blocks of weight less than 160 kg/m^2 LB > 160—loose blocks of weight greater than 160 kg/m^2 (roughly equivalent to > 100 mm thickness) LBG—grassed equivalent to > 100 mm thickness
OSA	Open stone asphalt
P	Piling
PC	Precast concrete slabs
R	Riprap R_x is equivalent to $D_{n50} = x$ mm
SA	Sand asphalt
SG	Sack gabions
SR	Soil reinforcement systems SR1—gravel fill 14–22 mm SR2—gravel fill 38 mm SR3—stone fill 150 mm SRG—grassed SRC—concrete filled

Use of granular filters and geotextiles

5. Use of granular filters and geotextiles

5.1. NEED FOR FILTERS

It has been found that failures of banks are often a result of inadequate design or construction of the underlayer, rather than being due to instabilities in the cover (or armour) layer. The cover layer provides the front line of defence against hydraulic attack and therefore needs to be sufficiently strong to withstand lift and drag forces, as well as wave impact and abrasion action in some cases. Nevertheless, it is at the interface between the cover layer and the underlying soil that some of the most critical flow and geotechnical conditions occur. Although they may not be very severe, they can sometimes become critical through being neglected or misunderstood at the design stage. These conditions are affected by the properties of the base soil and of the revetment, such as their relative permeabilities and particle size.

It was mentioned in Section 2.1 that an increase in the level of saturation of the material generally produces deterioration of soil strength, as the contact between the particles is reduced by the water enveloping them. Changes in the water content of the soil caused by seasonal variations or rapid rises in water level are particularly detrimental to the internal stability of banks, since these are stable for specified soil conditions; measures should be taken to minimise the effect of any departure from those conditions. In the presence of a cover layer, which will commonly have particles bigger than those of the base soil, there is a risk that fines from the soil will be washed through the larger voids of the cover layer (see Figure 5.1).

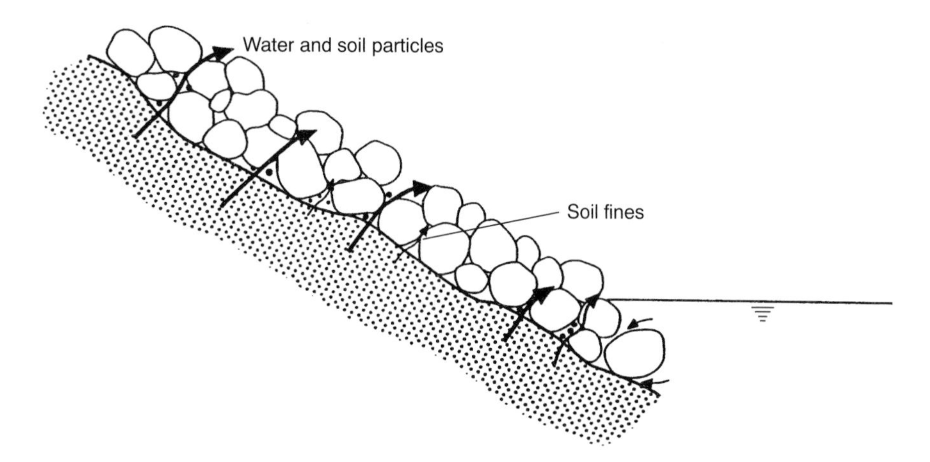

Figure 5.1. Example of soil bank with riprap cover layer and no filter

Underlayers are introduced into revetments to avoid the migration of soil particles and to allow, at the same time, the movement of water from the bank into the channel (and vice versa, in certain cases). Besides acting as filters, they can also fulfil other functions, such as separation between layers and regulation of the base soil for better placement of the cover layers. Underlayers can also be used as drainage media to provide a preferential path for water parallel to the cover layer. When the underlayer is designed for this function, which is more generally required for wave environments, it is important to ensure adequate outlet for the flow, either by provision of weepholes in impermeable revetments or by specifying sufficiently open revetment structures.

Filters can be made of two very different types of material (granular materials and/or geotextiles) which are described next. Whatever the type of filter, it is essential to ensure good contact with the underlying soil. Any pockets of air or obstacles between the filter and the base soil can jeopardise the desired filtering function: they help the establishment of preferential paths for water flow that will then bypass the filter. They can also result in differential settlement in the case of granular filters and flow-induced flapping in the case of geotextiles.

5.2. TYPES OF FILTER

Filters used in river engineering are generally either granular or made of synthetic materials, although composite types comprising the two materials may also be used. Both types can be effective at reducing the hydraulic gradients behind revetments by allowing the release of excessive water pressure without loss of the soil particles. Since their hydraulic responses can be made similar, the choice between granular and synthetic filters is often based on practical considerations, such as availability of materials and space, ease of construction, location, etc.

5.2.1. Granular filters

Granular filters have a long history of successful applications. Experience in their use has been complemented over the years by research into their properties and functioning, so that design principles are well established. This type of filter is formed, as the name indicates, by materials with a granular nature. The concept behind their use is that a gradual transition should be introduced between the base soil, which has on average smaller particle sizes, and the cover layer, which is made of much coarser material. Depending on the relative sizes of the particles in the two mediums, this transition may require more than one type of granular material. In fact, a granular filter usually needs to be formed by at least two layers (see Figure 5.2(a)), particularly if protecting very fine base soils.

In spite of the advantages mentioned above (used and proved solution), granular filters are difficult to install in some cases and today are often substituted by geotextiles. Examples of situations where granular filters are generally not advisable (for constructional, economical or hydraulic reasons) are as follows.

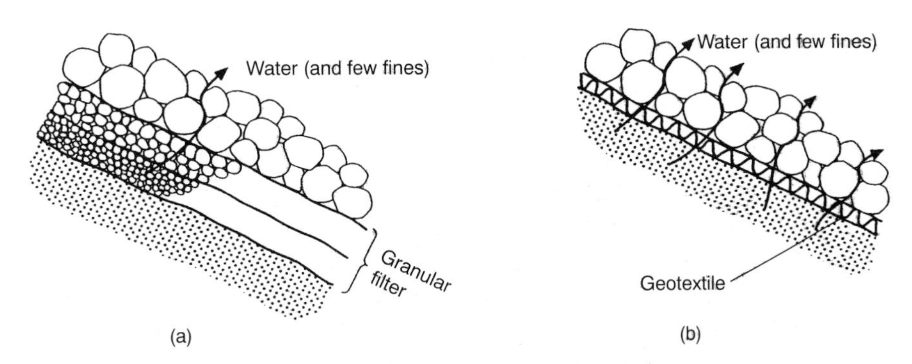

Figure 5.2. Example of soil banks with: (a) granular filter; (b) geotextile

- Construction under flowing water conditions (particularly in bed protection schemes without provision of cofferdams), where it may be difficult to ensure the stability of the various granular layers.
- Construction on banks with little available space, where the thickness required for the granular filter layers is not acceptable.
- Regions where granular materials with the necessary properties and quality are hard to obtain and/or control.
- High turbulence flows (see Section 2.4.1). Research on riprap revetments has shown that conventional granular filters can contribute to the destabilisation of the cover layer when the levels of turbulence are high. This is due to the strong uplift forces causing fine particles to move upwards into the voids, thereby reducing the contact between the stones.
- Clay soils. Granular filters are usually found to be more expensive and difficult to install than geotextiles since they normally require several layers.

5.2.2. *Geotextiles*

Geotextile is the generic name given to permeable textiles, meshes or nets used in contact with soil or rock. They are part of the wider category of geosynthetics, which also includes grids, geomembranes (impermeable liners) and geocomposites. Synthetic geotextiles should not be confused with biodegradable textiles, which are increasingly being used in river bed and bank protection schemes for environmental reasons. These latter have a very short life (of a few years) when compared with their synthetic counterparts and their properties are considerably different. Information on design of biodegradable textiles is not given in this book but can be found in Hemphill and Bramley (1989), for example.

There are three basic types of geotextile, with the great majority belonging to the first two categories (see Figure 5.3):

- woven, with fibres oriented at right angles and regular hole sizes
- nonwoven, formed by filaments or fibres randomly placed and with a wide range of hole sizes

Figure 5.3. Typical examples of nonwoven (left) and woven (right) geotextiles

- knitted, formed by fibres unable to move within the structure. This form of manufacture confers very high strength and flexibility, but such properties are not normally required in river applications.

Geotextiles also differ in the polymer material used for their fabrication. Polyester and polypropylene are the most common types, although polyethylene and other materials can also be found in a few cases. Both polyester and polypropylene have mechanical, filtration and chemical properties that render them suitable for erosion protection of river beds and banks. These properties can be summarised as follows.

- Mechanical (measured in terms of ultimate tensile strength, ultimate extensibility, creep characteristics)
  - ability to sustain loads during installation and service life
  - ability to resist damage during installation
  - abrasion resistance.
- Filtration
  - ability to filter soil.
- Chemical and biological
  - ability to resist ground environment
  - ability to resist ultra-violet (UV) and light.

The most important properties are summarised in Table 5.1. However, it is the view of some experts that nonwoven, needlepunched geotextiles are generally the most appropriate type when filtration is their main function. The reasoning behind this view is that, since strength (a distinguishing property of woven geotextiles) is not fully mobilised in the majority of cases, it is not necessary for

Table 5.1. General characteristics of geotextiles used in river engineering

<table>
<tr><th rowspan="2">Type</th><th rowspan="2" colspan="2">Description/appearance</th><th rowspan="2">Materials</th><th colspan="3">Properties</th></tr>
<tr><th>Mechanical</th><th>Filtration</th><th>Chemical</th></tr>
<tr><td rowspan="2">Woven</td><td rowspan="2" colspan="2">Orientation of fibres at right angles
Holes of uniform size
Appearance: textile-like (warp and weft)</td><td>Polyester (white)</td><td>Very high strength
100–1000 kN/m
(warp)</td><td>Pore size (O_{90})
70–300 μ =
= 0·07–0·3 mm</td><td>High UV resistance</td></tr>
<tr><td>Polypropylene
(usually black)</td><td>High strength
17–300 kN/m
(warp)</td><td>Pore size (O_{90})
200–550 μ =
= 0·2 to 0·55 mm</td><td>Very high UV resistance (suitable for tropical climates)</td></tr>
<tr><td rowspan="3">Nonwoven</td><td rowspan="2">Randomly laid filaments or fibres</td><td rowspan="2">Needlepunched
The filaments are agitated by barbed needles
Appearance: felt-like</td><td>Polyester (white)</td><td rowspan="2">Medium strength
6–100 kN/m
(warp)</td><td rowspan="2">Pore size (O_{90})
45–105 μ =
= 0·045 to
0·105 mm</td><td>High UV resistance</td></tr>
<tr><td>Polypropylene
(usually black)</td><td>Very high UV resistance (suitable for tropical climates)</td></tr>
<tr><td>Holes of irregular shape and sizes</td><td>Heatbonded
The filaments are heated between rollers until they partially melt together
Appearance: thin fabric, smooth to the touch</td><td colspan="4">Less common type—for information consult relevant manufacturers</td></tr>
</table>

geotextiles in river applications to have very high strength, but it is essential to ensure good filtration of the base soil. Woven textiles can be designed to perform very well as filters for a particular soil particle size since they have uniformly sized openings and can provide big porosities. However, nonwovens offer a wide range of opening sizes and can therefore be more effective as filters. They also have the ability to stretch much more than woven geotextiles without reaching breaking point; this is a useful property, for example in situations where the geotextile is placed behind large riprap and needs to stretch to achieve good contact between soil and stones.

There are other aspects related to geotextiles that need to be taken into account as they may greatly affect their performance. For example, it is important to ensure that geotextiles are covered (i.e. protected from light) not only during their design life but also while on site awaiting installation. This is a precaution against damage, especially from ultra-violet rays, which can considerably reduce the durability of the geotextile. A design life of 125 years is usually assumed for geotextiles in temperate climates; however, in tropical regions and particularly in marine environments, this value can be substantially reduced. Also, in environments where the ground or water pH is above 10 (e.g. next to setting concrete) polypropylene is preferred over polyester because of its higher resistance to chemical attack. Another aspect worth considering is the possible damage caused by dumping and spreading of large riprap or other bulky and angular revetments on a geotextile layer. It is obviously essential to protect the integrity of the membrane to ensure proper filter performance during its design life. The main disadvantages of geotextiles include a propensity for clogging or blocking, which can affect their filtering function, and insufficient knowledge of their long-term performance.

It is not within the scope of this chapter to provide fully comprehensive information on geotextiles, which constitute a very wide and complex field. The river engineer should bear in mind that specialist advice should be sought for non-standard applications. For an in-depth view of the subject it is recommended that the following publications be consulted: Rankilor (1981 and 1994) and Raymond and Giroud (1993).

5.3. DESIGN METHODS

The flowchart in Figure 5.4 presents the various steps necessary for filter design. The design of filters, irrespective of the type being considered, requires a knowledge of some geotechnical characteristics of the base soil (type of soil, grading curve). It is also necessary to have an idea of the type of revetment that will be used since its permeability needs to be taken into account. The next step involves the choice of filter type. As was mentioned above, criteria such as availability of space and materials, ease of construction, flow conditions and whether the revetment is to be built in the wet or in the dry will dictate the choice between a granular and a synthetic type of filter. Once the type has been established, the design methods will differ, as shown below.

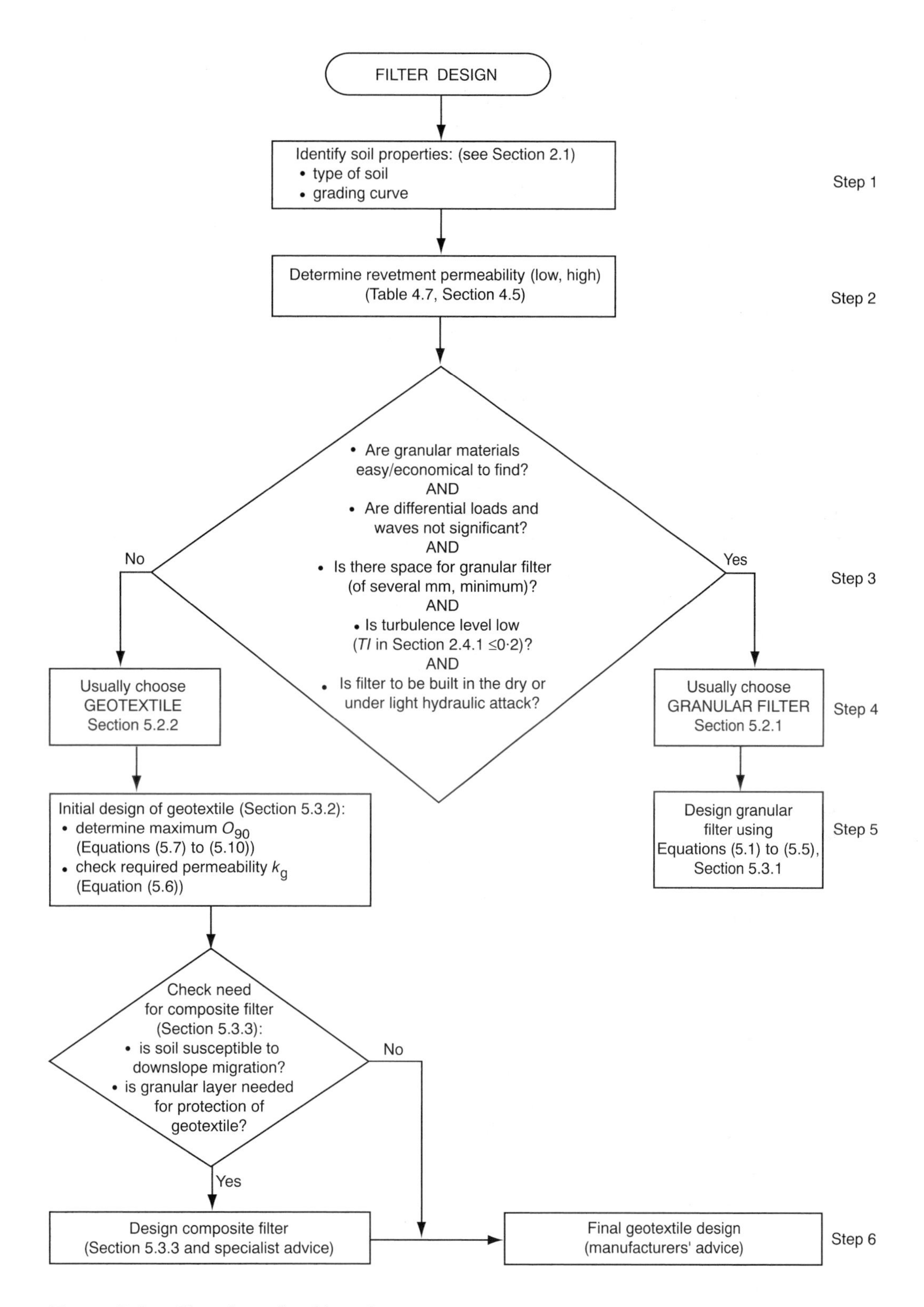

Figure 5.4. Flowchart for filter design

5.3.1. Design of granular filters

Granular filters are specified in terms of their grading curve and their thickness. The principle behind the design of filters is that it is necessary to assure stability of the cover layer and of all the successive layers of filter. This is more easily achieved when the grading curve of the filter is approximately parallel to that of the cover layer (this comment obviously applies only to granular types of revetment, such as riprap). When, as is common for base soils with fine particles, the filter needs to include several layers, each underlying layer will be taken as the filter and the layer above as the cover. The following criteria, based on studies carried out by several authors were derived from the well-known Terzaghi rules, and are applicable to the design of each layer, starting from the top:

$$D_{85f} \geq 0{\cdot}25 D_{15c} \tag{5.1}$$

$$D_{50f} \geq 0{\cdot}14 D_{50c} \tag{5.2}$$

$$D_{15f} \geq 0{\cdot}14 D_{15c} \tag{5.3}$$

$$D_{10f} > 0{\cdot}1 D_{60f} \tag{5.4}$$

where D_x is the particle size for which x% of the particles by weight are smaller, and the subscripts f and c denote filter and cover, respectively. In the above equations, the first three ensure stability of the cover layer and the last one the internal stability of the filter layer. The design proceeds until the sizes required for the filter layer match those of the base soil. However, it is also possible to start from the base soil and apply the above rules from the bottom to the top.

In the case of cover layers formed by concrete blocks, there is no standard procedure for the design of filters. One possible method is described next:

- determine the maximum size of cells or gaps between the blocks allowing for acceptable tolerances
- take this size as the D_{10} of the filter layer underneath the revetment
- determine D_{60} of the filter using Equation (5.4) and define an approximate grading curve
- check if characteristics of base soil match the above curve, if not apply Equations (5.1) to (5.4) until the requirements for the filter correspond to characteristics of the base soil.

Filters designed to have an additional drainage function need to have a considerably higher permeability than the base soil. Since this property is governed by the smaller particle sizes, the following criterion should be used for design:

$$D_{15f} > 5 D_{15b} \tag{5.5}$$

where subscript b refers to the base soil. It is also recommended that $D_{5f} > 0{\cdot}075$ mm to minimise the risk of blockage of the filter by fine particles.

Other design guidelines are as follows.

- The thickness of each filter layer should not be less than approximately 150 mm, less than D_{100} or less than 1·5 times D_{50}.
- If placement under water is feasible and the only satisfactory solution, the thickness of the layers should be increased by 50%.

5.3.2. *Design of geotextiles*

The design of geotextiles for use in river applications is based on the two main functions expected from them: permeability (the ability to convey flow either normal to their plane or along it) and retention (the ability to retain soil particles of specified size(s) while allowing flow of water). The filtering process can be described as follows: at the same time that the geotextile prevents bigger particles from passing through it, part of the small size fraction is allowed to migrate; the bigger particles therefore stay close to the geotextile and smaller ones will move to fill the gaps between them. It can be said that the role of the geotextile is to help the soil filter itself. This process is, however, only achieved when there is good contact between the geotextile and the soil.

In order to satisfy the first design criteria, permeability, it is necessary to ensure that the geotextile has higher permeability, k_g, than the underlying soil, k_s, and that head losses through the filter are within the specified limits (CUR, 1995):

$$k_g \geq M k_s \qquad (5.6)$$

where

M is a coefficient dependent on the type of geotextile

$M = 10$ for woven geotextiles
$M = 50$ for nonwoven geotextiles

k_s is the soil permeability. k_s is usually determined by laboratory testing but, in the absence of this, it can be estimated for sandy soils as:

$k_s = 0{\cdot}01 \times (D_{10})^2$, where D_{10} is in mm and k_s is in m/s.

Many different guidelines have been proposed to satisfy the retention design criterion, but they are generally all expressed in terms of the opening size of the geotextile that will retain the soil particles effectively. Research has demonstrated that it is the bigger holes in a geotextile which control the geotextile filtering function. For this reason it is common to choose O_{90} (and less frequently O_{95} and O_{98}) as the relevant parameter to define the porosity of a geotextile: this is the opening size of the geotextile corresponding to the diameter of the largest particles able to pass through it. In a similar way to soil grading curves, it is possible to define hole distribution curves for geotextiles: nonwoven geotextiles have a wide range of hole sizes while the openings in woven geotextiles are essentially uniform in size.

For design purposes, the O_{90} of the fabric is compared with the D_{90} of the underlying soil. It has been found that, contrary to what might be expected, the fabric openings can be substantially bigger than the soil particles before these will start to pass through the fabric. In granular soils this is due to 'bridging' over the openings that effectively increases the grain size; in cohesive soils it is a result of colloidal forces between particles. This phenomenon may, however, be counter-balanced by turbulence or strong wave action. The following general rules are given for design (Rankilor, 1997):

For woven geotextiles

$$O_{90} \leq 2{\cdot}5D_{90} \tag{5.7}$$

For nonwoven geotextiles

$$O_{90} \leq 5{\cdot}0D_{90} \tag{5.8}$$

Experience has shown that nonwoven fabrics can be more effective filters than woven fabrics, as they provide a wider range of pore sizes and can therefore cope better with uncertainties or variability in the soil characteristics. It is also apparent that the type of flow attack plays an important role in the design of geotextiles, which is not taken into account in Equations (5.7) and (5.8). The following guidelines are therefore given as rules of thumb to assist the designer:

$$O_{90} \leq 2D_{90} \quad \text{for rivers with limited wave action and normal turbulence levels} \tag{5.9}$$

$$O_{90} \leq D_{90} \quad \text{marine environments with breaking waves} \tag{5.10}$$

All the above criteria are based on geometric sandtightness, which can be difficult to meet in the case of fine soils. Recent work has been carried out in the Netherlands on the concept of hydrodynamic sandtightness. This approach, described in Breteler *et al.* (1995) is based on the observed fact that hydraulic loads are considerably reduced in the vicinity of the geotextile. The authors recommend its use for design of river bed and bank protection but suggest the application of traditional criteria for breaking waves and protection of bridge piers.

The guidance given in this book is suitable for the initial design of geotextiles. As pointed out above, this is very much a specialist area. The guidelines given here should be complemented with information obtained from other sources, particularly when the river engineer is involved in the design of large schemes or is dealing with unusual flow and soil conditions (see, for example, CUR, 1995, PIANC, 1987 and follow manufacturers' technical advice). For information on specification of geotextiles, refer to Appendix 5.

5.3.3. *Design of composite filters*

In some cases, the introduction of a granular layer between the cover layer and a geotextile filter is used for the following purposes:

- to reduce the hydraulic gradient within the base soil due to seepage
- to protect the geotextile during construction of cover layers formed by larger riprap
- to provide short term revetment protection in case of local damage to the cover layer

Some soils, generally in the sandy silt and fine sand range, are susceptible to downslope migration due to hydraulic loading. This process results from a momentary lifting of particles by hydraulic pressure, which separates them and dilates the soil. When reconsolidation occurs, it generates a gravity-induced downslope movement. To identify whether a soil is prone to this type of instability, the designer should carry out the following checks.

- Is a percentage of the soil particle sizes smaller than $60\,\mu$ and $D_{60}/D_{10} < 15$?
  OR
- Is more than half the soil particle sizes bigger than $20\,\mu$ but smaller than 0·1 mm?
  OR
- Is the plasticity index (defined as the difference between the liquid and plastic limits of fine soils which are determined from tests) $I_P < 0{\cdot}15$?

If the answer to any of these questions is 'yes', a granular underlayer should be incorporated between the cover layer and the geotextile.

It is recommended, however, that the use of granular layers underneath concrete block mattresses of low porosity and in highly turbulent flows be avoided, as they can promote uplift forces that the mattress may not be able to withstand. In these cases, the geotextile should be specified as thick and with an additional coarse fibre layer underneath (see PIANC, 1987 for guidance on design).

In these composite systems the geotextile is designed to function as a filter and therefore the granular layer does not need to fulfil that role. Its design is therefore based on ensuring that it has greater permeability than the base soil and the geotextile, and that it has a high weight per unit area. The thickness of this layer is typically between 100 mm and 500 mm, the upper limit applying to the most severe wave conditions.

Construction issues

6. *Construction issues*

There are a number of construction-related aspects that dictate not only the construction procedures to adopt, but also influence the choice of revetment. It has been pointed out in Chapter 3 that the design stages which the river engineer needs to go through involve the consideration of a multitude of different aspects if an integrated design is to be achieved. Construction issues are an important part of design that should be taken into account from the early stages. Knowledge of all the possible construction restraints is very much dependent on past site experience, but a few major aspects are described below to raise awareness of the subject.

6.1. AVAILABILITY AND SUPPLY OF MATERIALS

This is probably the first aspect to be considered. Availability of materials is essentially an economic consideration, since it is possible today to transport materials to any part of the world. It is linked, however, with the decision of most engineers and clients to specify the use of local materials or products easily obtainable wherever feasible, in order to minimise costs and environmental impact. For example, concrete block revetments have been used as substitutes to riprap in areas where stone is scarce, and in regions where vegetation covers are easily established soft revetments have grown in popularity. As well as availability, it is also important to ensure that an adequate supply and quality control of materials will be in place during construction. This is particularly relevant for large schemes, requiring substantial amounts of revetment or filter materials.

6.2. AVAILABILITY OF EQUIPMENT AND LABOUR

Whenever possible, it is advisable to rely on locally available machinery and on local labour skills. Traditional construction methods may, in some cases, provide a good alternative to more engineered solutions, provided that they are applied in conditions comparable to those of previous successful works. In situations where equipment (and also materials) have to be imported, costs related to import and export duties need to be taken into account, as do the inevitable time delays associated with importation.

6.3. ACCESS AND INFRASTRUCTURE

The methods of construction to adopt and the type of revetment are often dictated by the access and infrastructure available on site, or the potential for their development. Restricted land access will normally mean that construction will require waterborne equipment (barges, for larger quantities of material, or pontoons, for smaller schemes). On the other hand, where sufficient space is available on land, equipment will tend to be operated from the top of banks. Infrastructure facilities such as access roads may need to be purpose-built to guarantee adequate supply of materials, equipment and labour to the site.

6.4. ENVIRONMENTAL CONSIDERATIONS

Environmental aspects are an increasingly strong factor affecting the construction of bed and bank protection, and are legal requirements in the UK. Some of these aspects are outlined below.

- *Prevention of water pollution.* Although difficult to eliminate completely, the disturbance caused by soil excavation and consequent release of silt into the water stream needs to be minimised as this can affect the survival of aquatic species and the quality of the water for human consumption.
- *Prevention of ecological disturbance.* Construction should be carried out with care to minimise disturbance of nesting sites, and plant and animal habitats. This is related to the timing of construction work, which is addressed in Section 6.5.
- *Noise limitation and traffic restriction.* Noise due to construction should be within the limits considered acceptable for the workers, nearby populations and riverine animals. Restrictions to traffic to and from the site may also affect the construction method.

6.5. TIMING

One of the environmental aspects mentioned above regards the timing for execution of works. In order to minimise the impact on river species, it is best to avoid the invertebrate active period (in the UK from March to October); specific site conditions should be considered on an individual basis. However, the conditions most favourable for accuracy of installation of revetments correspond to this dry season, where water levels and flow velocities are low. This creates a conflict which is not always easily resolved.

Timing is a particularly relevant factor for construction in tidal conditions: work periods should be chosen so as to coincide with low water levels. Furthermore, phasing is an important consideration and work should be carried out, whenever possible, during low water periods of spring tides.

6.6. OPERATIONAL SITE CONDITIONS

These include:

- current, wave and wind conditions
- changes in the above due to tidal variation and seasonal influences (e.g. monsoons, flash floods)
- available water depth and manoeuvring space for barges when construction is carried out from the river
- for waterborne operations, sufficient space for the safe passage of other vessels
- visibility (above, and especially below water).

6.7. HEALTH AND SAFETY

Local and international regulations on health and safety of the workers and waterway users should be complied with during and after construction. Safety considerations may also assume an important role when choosing the type of revetment to adopt or the slopes at which they are installed in populated areas, since the risk of accidents will obviously need to be minimised.

From a chemical point of view, the great majority of revetments, and all of those dealt with in this manual, are safe because of their inert nature. However, it is important to point out that certain materials have been used, such as some industrial by-products and waste materials that, although adequate from a stability viewpoint, may release harmful components into the waterway. Examples of these materials are mining waste with high heavy metal contents and construction waste, such as rubble (particularly if it contains asbestos).

Stability of uncompleted works is another aspect that is disregarded in many cases, but of importance particularly when construction is being carried out in flowing water. Care should be taken not to exceed stable slope angles, for example when dumping riprap on banks; adequate protection of the bank toe should, if possible, be ensured prior to the rest of the revetment work being done since this is probably the most vulnerable area. Fixing of edges and anchoring are two other aspects that require careful execution in order to achieve correct levels of safety during construction and long-term integrity of the works.

Other legal requirements related to health issues, which were mentioned in Section 6.4 are to minimise pollution of the waterway and noise during construction.

6.8. TOLERANCES

Tolerances required for above and below water placement are necessarily different. In the first case they derive not only from requirements of stability and smoothness of underlayers but are also sometimes closely linked with aesthetics; the appearance of revetments (colour, texture, etc) is particularly important in urban areas. Underwater visual requirements are not relevant but smoothness of bed protection may be important to guarantee minimum navigation depths.

On banks, the cross-sectional tolerance of revetments is usually defined as the variation perpendicular to the slope. Uneven surfaces not only affect the stability of the revetment, but also increase the hydraulic roughness and therefore reduce the conveyance of the waterway. This applies both to the banks and the beds of rivers and channels.

6.9. CONSTRUCTION UNDER WATER

Construction of revetments under less than about 0·5 m of water (when placing by hand) or 1 m of water (when placing by machinery) in general does not present much increased difficulty when compared with dry construction. It can be carried out by land-based equipment and stability of the revetment materials is easily achieved in still or slow flowing water. When protecting from the top of the bank toes that lie under water, experience has shown that rounded shapes such as boulders or sausage gabions are easier to handle, since they can roll down slopes.

Underwater construction in depths greater than 1 m will usually require specific techniques to ensure correct positioning and fixing of the revetment. For example, granular filters tend to be substituted by geotextiles, and mattresses can be preferred over loose materials for ease and accuracy of installation. When concrete block mattresses are used in areas of high turbulence or fast flows, blinding of the blocks with gravel is not necessary, and may even be detrimental, as the gravel will be washed away. Furthermore, if no geotextile is present underneath, the stability of the revetment itself may be in question due to enhanced uplift forces.

Maintenance procedures

7. Maintenance procedures

To a lesser or greater extent, all types of revetment will require some maintenance during their serviceable lives. In broad terms, solutions consisting of bioengineering or composite types involving vegetation and structural materials are generally more heavily dependent on frequent maintenance if they are to retain their properties and fulfil their function (see Table 4.7 in Section 4.5). Particular needs of these revetments include accessibility for cutting machinery and for workers, and appropriate choice of the maintenance periods in order to maximize the effectiveness of the operations and minimise ecological impact. Consultation of specialist publications such as Morgan *et al.* (1998) and Hemphill and Bramley (1989) is recommended.

7.1. MAINTENANCE PROGRAMME

Experience has shown that many well designed and built structural revetments (e.g. concrete linings, riprap or gabions) can have a very long life, of several decades, only needing minimal maintenance. Most of this maintenance will involve inspection and monitoring rather than routine repairs. In fact, maintenance programmes should comprise a range of activities with more or less emphasis on certain items, depending on the magnitude of the revetment scheme and the type of solution. According to guidelines set up by PIANC (1987), a maintenance programme should include the items detailed below.

7.1.1. *Record of the watercourse and revetment characteristics*

This may take the form of a database, which should be regularly updated. Relevant information concerning the watercourse may consist of: length, width, depth, flow discharge, average flow velocity, waves, existence or absence of tidal movement and navigation, water quality. Information concerning the adjacent land (e.g. vegetation, land use, groundwater) should also be included, as well as photographs, drawings and details of typical cross-sections. These should give: the date of installation of the revetment, cover layer, filters and subsoil data, and description of previous revetments.

7.1.2. *Establishment of acceptable standards*

These may or may not be similar to those in place during construction, and will vary depending on the location of the works (particularly if they are under water) and on the severity of the hydraulic loads. It is good practice to set standards so that they relate to the various degrees of maintenance possible. Requirements can therefore be established which, if satisfied, will mean the continuation of the existing inspection and monitoring plans; when not satisfied, different levels of requirements may mean a stricter inspection programme or the repair and eventual replacement of the revetment.

7.1.3. *Assessment of the state of the revetment*

The comparison between the current state and the original condition can be carried out in three ways: visual inspection, detailed measurements and monitoring using equipment installed during or after construction (this latter method is usually only applicable to very large schemes of vital importance). The maintenance programme should set guidelines for increasing the frequency and detail of inspection when the minimum acceptable standards are close to being breached. Deterioration of the revetment (for example the height of a bank or the thickness of the cover layer) can be monitored with the aid of graphs showing evolution in time. Inspection of underwater revetments requires adequate specification and is carried out by divers and radio equipment. It is important to appreciate that the work of divers is often hindered by very low visibility. Good paper and photographic records and drawings of the built revetments therefore assume particular importance.

7.1.4. *Planning and execution*

These aspects are ruled by financial constraints as well as operational limitations such as the availability of space and access for execution of the work. Guidelines should be given to help the engineer to decide on whether to maintain the revetment, replace it with another, or choose a different strategy to solve the problem. Assistance on problem identification and various alternative strategies in use in the UK can be found in Morgan *et al.* (1998) and Thorne *et al.* (1996). As mentioned in Section 6.5, the time and execution of non-urgent maintenance work should, if possible, be set to coincide with the less active fauna and flora seasonal periods, which in the UK extend from October to March.

Worked examples

8. Worked examples

8.1. WORKED EXAMPLE 1

The banks of a river from 5 m downstream of a gated weir have been left unprotected against erosion after construction of the weir. They are experiencing severe erosion, which is also endangering a nearby public footpath. The reach in need of protection has a very mild longitudinal slope, is exposed to wind and comprises a sharp bend (radius of about 1000 m) at the end of a 300 m long straight stretch. The flow velocity in the river through a gate close to one of the banks is estimated to be 3·5 m/s, the water depth is 2·5 m and the surface water width is 50 m for the required design conditions. There are also signs of formation of scour holes in the river bed (water surface is quite irregular) close to the weir where the flow appears quite turbulent. The weir has a 5 m long concrete apron on the downstream side. The soil in the banks and bed is a silty sand with the grading curve shown in Figure 8.1.

Design a suitable revetment. The design should follow the stages described in Section 3.3.

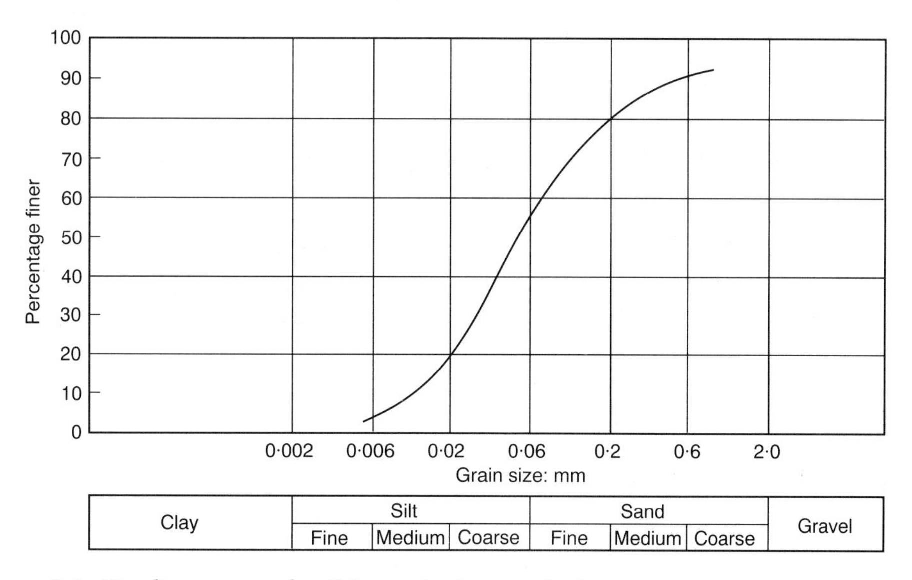

Figure 8.1. Grading curve of soil in worked example 1

8.1.1. Conceptual design — Section 3.3.1, flowchart of Figure 3.1

The situation described above involves an assessment of the stability of the river bed and banks. The information given shows instability in unprotected river banks with no previous history of protection. Due to the proximity of the weir, the causes of failure are likely to be high turbulence in the flow, aggravated by the presence of a bend. Strong winds in this exposed area may also produce waves that can destabilise the banks. Without further assessment, the main consequence of erosion of the river banks is the eventual loss of the nearby footpath and possibly also of valuable adjacent land. As for the erosion of the river bed, this can eventually lead to the undermining of the concrete apron and to the collapse of the structure. The steps in the flowchart of Figure 3.1 involving the setting of objectives and strategies will not be considered in detail in this example, but the designer should consider the available access for construction.

It is necessary next to determine the type of hydraulic attack to which the reach is subjected. The mean flow velocity has been previously estimated to be 3·5 m/s but the likely wave height still needs to be determined.

The stretch of river under consideration is not subject to significant boat traffic. Due to the exposed nature of the site, the maximum waves are likely to be wind-induced waves (see Section 2.3.2), although turbulence can also generate significant surface waves. It can be seen in Figure 2.12 of Section 2.3.2, that case (c) depicts a situation that is similar to the one under analysis. In view of the lack of specific wind data, take U_{10} as 25 m/s (for exposed site):

$$F = (3L_w + 67B)/40 = (3 \times 300 + 67 \times 50)/40 = 106\,\text{m}$$

and from Equation (2.9)

$$H_s = 0{\cdot}00354 \times (U_{10}^2/g)^{0{\cdot}58} F^{0{\cdot}42} = 0{\cdot}00354 \times (25^2/9{\cdot}81)^{0{\cdot}58} \times 106^{0{\cdot}42} = 0{\cdot}280\,\text{m}$$

Since the flow velocity $U = 3{\cdot}5$ m/s $> 2{\cdot}5$ m/s and the design wave height $H = 0{\cdot}280$ m $> 0{\cdot}15$ m, and the site is near a hydraulic structure, a structural or biotechnical solution is recommended.

8.1.2. Outline design — Section 3.3.2, flowchart of Figure 3.2

More design data should be obtained during this stage. Factors that may affect the choice of solution are the probable need for some underwater construction and limited access available along the bank. Since the weir is gated, flow can, in principle, be considerably reduced for limited periods of time to facilitate construction.

The predominant loads are:

- currents, $U = 3{\cdot}5$ m/s (Section 2.3.1)
- wind waves, $H_s = 0{\cdot}280$ m (Section 2.3.2)
- high turbulence (Section 2.4.1)

  In Table 2.6 it can be seen that the present situation corresponds to very high turbulence, $TI = 0{\cdot}60$.

- flow around a bend (Section 2.4.2)
  Calculate ratio $R/W = 1000/50 = 20$. This ratio is smaller than 26, so the bend will have a marked effect on the velocity patterns. However, the turbulence created by the gated weir will be much greater than that due to the bend, so the effect of the latter can be neglected.

As a first approach, and according to the flowchart of Figure 5.4 (up to Step 4) in Chapter 5, a filter layer is likely to be required to protect the banks and part of the bed of the river. Since the turbulence of the flow is high and the space available along the outer bank of the bend is limited due to the footpath, a synthetic filter (geotextile) is advisable. Study of Chapter 6 indicates that installation of a geotextile is feasible.

Alternative solutions should be compared next. The governing flow condition is current attack with high turbulence and the channel is non-tidal. Table 3.2 in Section 3.2 indicates the following possible types of revetment:

- riprap
- grouted stone
- box gabions
- gabion mattresses
- cabled blocks
- bituminous materials
- concrete
- piling.

For detailed information on the above systems, Chapter 4 and the Data Sheets of Appendix 1 should be consulted at this stage. Table 4.8 in Section 4.5 gives an indication of the suitability of the various types of revetment for the current and wave flow values but it should be borne in mind that the table was devised based on normal to medium turbulence levels. Although sheet piling is an option for the protection of the banks, it will not be considered in this example because it is a form of vertical bank protection, and not strictly covered in this book. Table 4.8 shows that geomats with good grass cover, soil reinforcement systems with 150 mm stone fill, gabion mattresses 0·15 to 0·17 m thick and riprap of 500 and 600 mm are possible solutions. Further conclusions can be taken from Table 4.7 in Section 4.5 regarding the suitability of the above revetments. For example, Table 4.7 points out that geomats with grass cover and soil reinforcement systems are only generally viable on upper bank reaches.

For the protection of the main channel, the designer can therefore choose, for example, between riprap and gabion mattresses. Appendix 1 should be consulted to obtain more information on proprietary types of gabion mattresses. Assuming that in this particular case riprap is preferable to gabion mattresses and that the necessary large stone is locally available, a decision could be made to adopt riprap as the main protection of the bed and banks of the river; grassed geomats can be used on the upper banks.

8.1.3. *Detailed design — Section 3.3.3, flowchart of Figure 3.3*

Calculate stable stone size of riprap.

(a) Wind waves — flowchart of Figure 2.13, Section 2.3.2 and calculations under 'Conceptual design' (Section 8.1.1)

$F = 106\,\mathrm{m}$
$H_s = 0{\cdot}280\,\mathrm{m}$
Using Equation (2.10)
$T_z = 0{\cdot}581\,(FU_{10}^2/g^3)^{0{\cdot}25} = 0{\cdot}581\,(106 \times 25^2/9{\cdot}81^3)^{0{\cdot}25} = 1{\cdot}68\,\mathrm{s}$
Using Equation (2.11)
$H_i = 1{\cdot}3H_s = 1{\cdot}3 \times 0{\cdot}280 = 0{\cdot}364\,\mathrm{m}$

Take $s = 2{\cdot}65$ and use Equations (2.12) and (2.13)

$D_{n50} = 0{\cdot}34\,[H_i/(s-1)]I_r^{0{\cdot}5}$
$I_r = (\tan\alpha)/[(2\pi H_i)/(1{\cdot}3gT_z^{\,2})]^{0{\cdot}5}$

Assume bank slopes will be set at $1V{:}2H$, i.e. $\tan\alpha = 0{\cdot}5$ and $\alpha = 26{\cdot}6°$

$I_r = 0{\cdot}5/[(2\pi \times 0{\cdot}364)/(1{\cdot}3 \times 9{\cdot}81 \times 1{\cdot}68^2)]^{0{\cdot}5} = 1{\cdot}986$
$D_{n50} = 0{\cdot}106\,\mathrm{m}$

(b) Turbulence — Section 2.4.1

Note that there is no need to design for flow in bends in this case because of the high levels of turbulence. Use the three equations recommended.

Escarameia and May's equation

- Bed and bank protection
  Equation (2.17)

  $D_{n50} = C(U_b^2)/[2g(s-1)]$

  From Table 2.6 take $TI = 0{\cdot}60$
  From Table 2.7 $C = 12{\cdot}3\,TI - 0{\cdot}20 = 7{\cdot}18$

  Due to lack of data assume $U \cong U_d$ and
  $U_b = (-1{\cdot}48TI + 1{\cdot}36)\,U_d = (-1{\cdot}48 \times 0{\cdot}6 + 1{\cdot}36) \times 3{\cdot}5 = 1{\cdot}65\,\mathrm{m/s}$
  $D_{n50} = 7{\cdot}18 \times (1{\cdot}652^2)/(2 \times 9{\cdot}81 \times 1{\cdot}65) = 0{\cdot}605\,\mathrm{m}$

Pilarczyk's equation

Take $K_T = 2$

- Bed protection
  Use Equations (2.19) and (2.20) and Table 2.8

  $D_{n50} = [\phi/(s-1)] \times [0{\cdot}035/\psi] \times K_T \times K_s \times [U_d^2/(2g)] \times [D_{n50}/y]^{0{\cdot}2}$

  For bed protection, $K_s = 1$, and therefore

$D_{n50} = [0{\cdot}75/(2{\cdot}65-1)] \times [0{\cdot}035/0{\cdot}035] \times 2 \times [3{\cdot}5^2/(2 \times 9{\cdot}81)] \times [D_{n50}/2{\cdot}5]^{0{\cdot}2}$

Start iterative procedure to find D_{n50} by first assuming that $D_{n50} = 0{\cdot}6$ m. The left- and right-hand sides of the above equation give the same value for $D_{n50} = 0{\cdot}4$ m.

- Bank protection
  Calculate $K_s = k_d$ using Equation (2.21); take angle of internal friction of riprap ϕ equal to 40°
  $K_s = (\cos\alpha) \times [1 - (\tan\alpha/(\tan\phi))^2]^{0{\cdot}5}$
  $K_s = (\cos 26{\cdot}6°) \times \{1 - [\tan 26{\cdot}6°/(\tan 40°)]^2\}^{0{\cdot}5} = 0{\cdot}718$
  $D_{n50} = 0{\cdot}547$ m

Maynord's equation

- Bed protection
  Assume angular rock
  Equation (2.22)

  $$D_{30} = S_f\, C_s\, C_v\, C_T\, y\, \{[1/(s-1)]^{0{\cdot}5} \times [U_d/(K_1\, gy)^{0{\cdot}5}]\}^{2{\cdot}5}$$
  $$D_{30} = 1{\cdot}5 \times 0{\cdot}3 \times 1{\cdot}25 \times 1{\cdot}0 \times 2{\cdot}5 \times \{[1/(2{\cdot}65 - 1)]^{0{\cdot}5} \times (3{\cdot}5/(1 \times 9{\cdot}81 \times 2{\cdot}5)^{0{\cdot}5}]\}^{2{\cdot}5}$$
  $$= 0{\cdot}316\text{ m}$$

  Convert D_{30} into D_{n50} using Equation (4.5) in Section 4.1.1:

  $$D_{30} = 0{\cdot}70 D_{50}$$
  $$D_{50} = 0{\cdot}451\text{ m}$$

- Bank protection
  Calculate K_l
  $$K_l = -0{\cdot}672 + 1{\cdot}492\cot\alpha - 0{\cdot}449\cot^2\alpha + 0{\cdot}045\cot^3\alpha = 0{\cdot}876$$

  $$D_{30} = 1{\cdot}5 \times 0{\cdot}3 \times 1{\cdot}25 \times 1{\cdot}0 \times 2{\cdot}5 \times \{(1/1{\cdot}65)^{0{\cdot}5} \times [3{\cdot}5/(0{\cdot}876 \times 9{\cdot}81 \times 2{\cdot}5)^{0{\cdot}5}]\}^{2{\cdot}5} = 0{\cdot}372\text{ m}$$
  $$D_{50} = 0{\cdot}531\text{ m.}$$

From the above results of the three different equations choose for the bed and the banks

$$D_{n50} = 0{\cdot}6\text{ m} = 600\text{ mm, which corresponds to } W_{50} = 0{\cdot}6^3 \times 2650 = 572\text{ kg.}$$

Determine the layer thickness from Table 4.3, Section 4.1.1:

thickness $> 2 \times D_{n50} = 2 \times 0{\cdot}6 = 1{\cdot}2$ m

Characteristics of the grading curve:

$$D_{85}/D_{15} = 1{\cdot}5 \text{ to } 2{\cdot}5$$
$$W_{85}/W_{15} = 3{\cdot}4 \text{ to } 16{\cdot}0$$

For the specification of the riprap refer to Appendix 2. Some classes are given in this appendix to assist in the specification of riprap. However, since the size required for stability is larger than that of any of the given classes, these cannot be used. Specify other characteristics, such as density, colour and stone integrity, using the appendix.

(c) Design of the filter layer

From the flowchart of Figure 5.4 in Section 5.3 and Table 5.1 in Section 5.2, select nonwoven needlepunched geotextile. This type is suitable for the present case for the following reasons:

- fairly limited information is available about the soil material and about the likely hydraulic gradients
- the soil is well graded and therefore the geotextile needs to be able to filter a wide range of particle sizes
- the riprap size is quite large and the risk of damage to the geotextile should be minimised by choosing a type that can be stretched to maintain good contact with the stones.

Characteristics of the soil:

Type: granular soil, silty sand and well graded. From Table 2.1 in Section 2.1.1 the soil is expected to have fair to poor drainage characteristics.

From Figure 8.1:
$D_{10} = 0{\cdot}01$ mm
$D_{50} = 0{\cdot}05$ mm
$D_{60} = 0{\cdot}07$ mm
$D_{90} = 0{\cdot}6$ mm

For the determination of the O_{90} of the nonwoven geotextile, and because turbulence is the governing flow condition, it is recommended that Equation (5.10) in Section 5.3.2 be used rather than Equation (5.8), which would normally be recommended:

$$O_{90} \leq D_{90} = 0{\cdot}6 \text{ mm}$$

Table 5.1 in Section 5.2.2 shows that most nonwovens have O_{90} of less than 0·105 mm and therefore meet the above criterion. Therefore a correct design can be obtained with any nonwoven geotextile, preferably with O_{90} in the upper range.

Use Equation (5.6) in Section 5.3.2 to calculate required permeability k_g of the geotextile:

$$k_g \geq 50k_s$$

Estimate $k_s = 0{\cdot}01 \times 0{\cdot}01^2 = 1{\cdot}0 \times 10^{-6}$ m/s

$$k_g \geq 5 \times 10^{-5} \text{ m/s}$$

Determine susceptibility to downslope migration (Section 5.3.3):

Part of the soil is smaller than 0·06 mm and $D_{60}/D_{10} = 0{\cdot}07/0{\cdot}01 = 7 < 15$. A granular layer between the geotextile and the riprap would, in principle, be required. However, because the turbulence of the flow is very high, fine granular materials should be avoided as they can increase the instability of the cover layer. A thick geotextile should therefore be specified according to Appendix 5, with an additional fibre layer underneath designed following guidelines given in PIANC (1987). The final design of the geotextile should be discussed with manufacturers.

(d) Design of the upper bank protection

Grassed geomats can be used to protect the upper banks which are not regularly subjected to hydraulic loads (Section 4.4.3). A suitable proprietary system can be chosen from the data sheets in Appendix 1.

(e) Edge and toe detailing

Due to the presence of waves, the height of wave run-up h should be taken into account when deciding the top level of the protection (Section 2.3.2 and Appendix 6):

for riprap, $h = 2H = 2 \times 0{\cdot}28\,\text{m} = 0{\cdot}56\,\text{m}$

Due to the turbulence in the area, it is recommended that this value be increased to 0·75 m and that the banks be lined with riprap 0·75 m above design water level. Above this level, the banks can be protected with grassed geomats.

This protection system should be extended downstream of the bend to 1·5 times the flow width, or $1{\cdot}5 \times 50 = 75\,\text{m}$. Upstream of the bend the protection should be continuous to the weir.

For the specification of the riprap and the geotextile, see Appendices 2 and 5.

8.2. WORKED EXAMPLE 2

A problem has occurred in an artificial channel in a tropical East Asian country, where the lining has locally collapsed along a 400 m long reach of the channel. Upstream and downstream of this reach, the existing bank protection (a type of proprietary concrete block mat no longer being manufactured) appears in satisfactory condition. The channel is not navigable and has a trapezoidal cross-section with the following dimensions: base width of 28 m, side slopes of $1V{:}1{\cdot}5H$ and surface water width (for mean water level, MWL) of 40 m. The reach under consideration is in a fairly sheltered area and is tidal: high and low water levels are 1·5 m above and below MWL, respectively. At MWL the water depth is 4 m and the flow velocity is equal to 2·5 m/s. The soil forming the banks and bed of the channel is a poorly graded fine to medium sand, with the grading curve shown in Figure 8.2.

The design will follow the procedure laid out in Section 3.3.

8.2.1. *Conceptual design — Section 3.3.1, flowchart of Figure 3.1.*

The present case involves the redesign of damaged works. It is believed that the damage that has occurred to the lining of the bank is likely to have been due to deficient installation of the mattress in a localised area. Possible consequences of allowing the damage to continue are progressive scour and general bank instability, as well as negative visual impact.

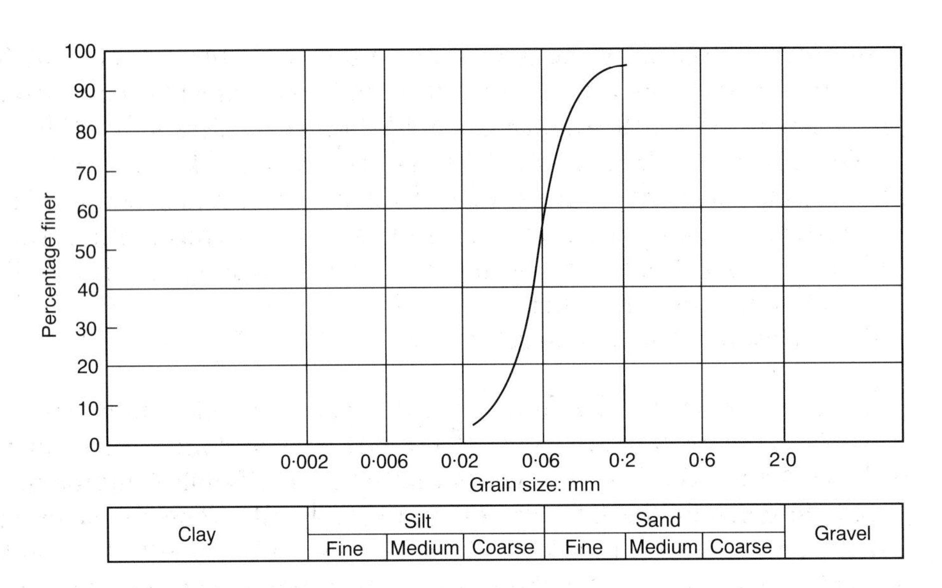

Figure 8.2. Grading curve of soil in worked example 2

The mean flow velocity in the channel is estimated to be 2·5 m/s. It is necessary to check the height of wind-generated waves. In view of the lack of specific data on wind velocities, U_{10} can be taken as 18 m/s (sheltered site). Consider cases (a) and (b) of Figure 2.12 (Section 2.3.2) to determine the maximum fetch F. Case (a) gives $F = B = 40$ m, whereas case (b) applies to situations where $L_w > 20B$. In the present case, $L_w = 400$, i.e. $< 20 \times 40 = 800$. The fetch can therefore be taken as equal to 40 m. The wind wave height is then given by:

$$H_s = 0{\cdot}00354 \times (U_{10}^2/g)^{0{\cdot}58} F^{0{\cdot}42} = 0{\cdot}00354 \times (18^2/9{\cdot}81)^{0{\cdot}58} \times 40^{0{\cdot}42} = 0{\cdot}13\,\text{m}.$$

The combination of $U \geq 2{\cdot}5$ m/s and waves points to the choice of a biotechnical or structural solution. For this particular situation, because the existing lining consists of concrete blocks, it is reasonable to choose concrete blocks for the remedial works.

8.2.2. Outline design — Section 3.3.2, flowchart of Figure 3.2.

More design data should be obtained at this stage. The main problems are the construction of the revetment and toe protection below 4 m of water (this can be minimised by timing some of the construction to coincide with low tides).

The predominant loads are:

- current attack, $U = 2{\cdot}5$ m/s (Section 2.3.1)
- wind waves, $H = 0{\cdot}13$ m (Section 2.3.2)
- tidal flows (drawdown) (Section 2.3.3).

In order to keep the same type of lining present in the channel, and in view of the fairly severe flow attack, ecological and environmental considerations will play a small role when compared with stability considerations.

With regard to the need for filter layers (Figure 5.4, Chapter 5), if cellular concrete blocks are used, the revetment will have an average permeability. Waves and differential loads are important in this case and construction will probably need to be carried out under water, so a filter consisting of a geotextile is more suitable than a granular filter.

Comparison of alternative solutions

- Table 3.1 in Section 3.2 shows that the type of attack is moderate.
- Table 3.2 in Section 3.2 shows that concrete blocks are suitable in principle.
- Table 4.7 in Section 4.5 indicates that both the flexibility and permeability of cellular concrete blocks are medium; they require low maintenance and can be built above and below water.
- Table 4.8 in Section 4.5 shows that cabled blocks can withstand mean flow velocities of the order of 4·5 m/s, with wave heights below 1·5 m.

From the above, cellular concrete blocks can be chosen, and these could be grassed above mean water level to enhance their appearance.

8.2.3. *Detailed design — Section 3.3.3, flowchart of Figure 3.3*

Determination of loads

(a) Wind waves: $H = 0{\cdot}13$ m; flowchart of Figure 2.13, Section 2.3.2 and calculations under 'Conceptual design' (Section 8.2.1)

$$F = 40\,\text{m}$$
$$H_s = 0{\cdot}130\,\text{m}$$
$$T_z = 0{\cdot}581\,(FU_{10}^2/g^3)^{0{\cdot}25} = 1{\cdot}12\,\text{s}$$
$$H_i = 1{\cdot}3 \times 0{\cdot}13 = 0{\cdot}17\,\text{m}$$

Use Equation (2.14) for cabled concrete blocks

$$D = (GH_iI_r^{0{\cdot}5})/[(s-1)(\cos\alpha)]$$
$$I_r = (\tan\alpha)/[(2\pi H_i)/(1{\cdot}3gT_z^2)]^{0{\cdot}5}$$

The bank slopes are set at $1V{:}1{\cdot}5H$, i.e. $\tan\alpha = \tan 33{\cdot}7° = 0{\cdot}667$

$$I_r = 2{\cdot}576$$

Take $s = 2{\cdot}3$ and $G = 0{\cdot}19$ (cabled blocks) in Equation (2.14)

$$D = [0{\cdot}19 \times 0{\cdot}17 \times (2{\cdot}576^{0{\cdot}5})]/(1{\cdot}3 \times \cos 33{\cdot}7°) = 0{\cdot}048\,\text{m}$$

The above thickness is for solid blocks; assuming an open area of 25%, the equivalent thickness is 0·064 m.

(b) Current flow, $U = 2{\cdot}5$ m/s

Use formulae in Chapter 4, Section 4.3.2:

Pilarczyk's equation

Equations (4.15) and (4.17)

$$D = 0{\cdot}026\ U_d^2/\{(1 - n)(s - 1)\ K_s\ [\log\ (12y/D)]^2\}$$
$$K_s = k_d = \cos\alpha\ [1 - (\tan\alpha/\tan\phi)]^{0{\cdot}5}$$

Assume $\phi = 35°$

$$K_s = \cos 33{\cdot}7°\ [1 - (\tan 33{\cdot}7°/\tan 35°)^2]^{0{\cdot}5} = 0{\cdot}253$$

Assume $n = 0{\cdot}25$ and $s = 2{\cdot}3$, and for first approximation in the iterative process assume $D = 0{\cdot}1$ m:

$$D = 0{\cdot}026 \times 2{\cdot}5^2/\{(1 - 0{\cdot}25)\ (2{\cdot}3 - 1) \times 0{\cdot}253\ [\log_{10}\ (12 \times 4/D)]^2\}$$

An agreement between the right- and left-hand sides of the equation is found for $D = 0{\cdot}087$ m, which is the thickness of blocks with 25% open area necessary for stability under current attack.

For the combined load of wind waves and currents, add the two sizes calculated in (a) and (b) to obtain the stable size $D = 0{\cdot}150$ m.

Table 4.8 in Section 4.5 indicates that concrete blocks with weight $< 250\ \text{kg/m}^2$ should be appropriate. In the data sheets presented in Appendix 1 it can be seen that, for example, the proprietary systems Armorflex 220 or Dycel 150 should provide the necessary protection. However, the designer should seek more detailed information from block manufacturers. If the open area of the blocks adopted differs by, say, more than 5% from the value assumed earlier, the design should be revised to calculate the required new block thickness.

(c) Design of the filter layer

From the flowchart of Figure 5.4 in Section 5.3 and Table 5.1 in Section 5.2, select woven geotextile. This type is suitable for the present case for the following reasons:

- the filter has to be installed under water
- the soil is poorly graded and therefore the geotextile does not need to be able to filter a wide range of particle sizes.

Characteristics of the soil:

Type: granular soil, fine to medium sand and poorly graded. From Table 2.1 in Section 2.1.1 the soil is expected to have fair drainage characteristics

$D_{10} = 0{\cdot}03$ mm
$D_{50} = 0{\cdot}06$ mm
$D_{60} = 0{\cdot}065$ mm
$D_{90} = 0{\cdot}1$ mm

Use Equation (5.7) to determine O_{90} for woven geotextile

$$O_{90} \leq 2{\cdot}5 D_{90}$$
$$O_{90} \leq 2{\cdot}5 \times 0{\cdot}1 = 0{\cdot}25\ \text{mm}$$

Table 5.1 in Section 5.2 shows that, for example, a woven geotextile with O_{90} equal to 0·2 or 0·25 mm meet this criterion; since the scheme is in a tropical climate, a polypropylene geotextile is recommended to minimise UV damage.

Use Equation (5.6) in Section 5.3.2 to calculate required permeability k_g of the geotextile:

$k_g \geq 10k_s$

Estimate $k_s = 0{\cdot}01 \times 0{\cdot}03^2 = 9 \times 10^{-6}$ m/s

$k_g \geq 9 \times 10^{-5}$ m/s

According to the rule of thumb presented in Section 2.3.3, in tidal flow conditions it is advisable that the permeability of the geotextile is at least 20 times greater than that of the soil, i.e. $k_g \geq 1{\cdot}8 \times 10^{-4}$ m/s.

Determine susceptibility to downslope migration (Section 5.3.3):

Part of the soil is smaller than 0·06 mm and $D_{60}/D_{10} = 0{\cdot}065/0{\cdot}03 = 2{\cdot}2 < 15$. A granular layer between the geotextile and the cover layer is therefore required. This granular layer needs to be more permeable than the geotextile, i.e. $k_{\text{granular layer}} > 1{\cdot}8 \times 10^{-4}$ m/s. $D_{10\ \text{granular layer}} = (k_{\text{granular layer}}/0{\cdot}01)^{0{\cdot}5} = 0{\cdot}13$ mm, i.e. fine sand. The thickness of this layer can be taken, in first approximation, as 100 mm.

The final design of the geotextile should be discussed with geotextile manufacturers. It is important to bear in mind that the life of geotextiles can be substantially reduced in tropical countries, when compared with other climates.

(d) Design of the upper bank protection

The cells of the concrete blocks in the upper banks above mean water level can be filled with top soil and seeded with natural vegetation.

(e) Edge and toe detailing

Due to the presence of waves, the height of wave run-up h should be taken into account when deciding the top level of the protection (Section 2.3.2 and Appendix 6)

For concrete blocks $h = 4H = 4 \times 0{\cdot}13\,\text{m} = 0{\cdot}52\,\text{m}$.

It is advisable that the whole bank be lined with concrete blocks up to the top; as previously mentioned, above mean water level vegetation can be encouraged to grow in the cells of the blocks.

It is recommended that the concrete block mattress be extended on the bed of the channel well beyond the toe of the bank.

For the specification of the concrete blocks and of the geotextile see Appendices 4 and 5.

Appendix I.
Details of river revetment systems

Table A1.1 Non-proprietary revetments

	Brief description
Rock	
Riprap	Loose stone randomly placed on more than one layer
Block stone	Fairly square stone units, typically over 1000 kg in weight
Hand pitched stone	Stone placed regularly by hand in a single layer
Grouted stone (see also bitumen-bound materials)	Riprap or hand pitched stone bound with cement or bitumen-based mortar
Concrete	
In-situ concrete lining	Solid and cellular concrete slabs
Pre-cast slabs	Regularly shaped concrete units; solid and more slender than concrete blocks
Bitumen-bound materials	
Permeable:	
Open stone asphalt and stone asphalt mats	Mixture of stone and mastic with an open structure; suitable for river bed and bank protection
Sand asphalt	Bitumen-coated sand; used as a revetment mainly above water and as a filter layer
Impermeable:	
Dense stone asphalt	Mixture of 50–70% stone with mastic; can be applied underwater
Asphalt concrete	Mixture of stone or gravel, sand and bitumen; for use mainly above water level
Mastic	Very flexible bitumen, sand and filler mix; used in impermeable layers and toe protection
Asphalt grouted stone	Mixture of mastic and stone filling the interstices of the cover layer
Flexible forms	
Filled sacks	Hessian or woven synthetic bags filled with concrete, sand or other mixes; usually forming near-vertical walls

Table A1.2 Proprietary revetment systems

Proprietary name	Manufacturer/supplier*	Brief description
Gabions		
Box gabions		
Maccaferri	Maccaferri Ltd	Double twisted wire mesh boxes with stone fill
Weldmesh	Weldmesh Land Reinforcement/Tinsley Wire (Wigan)	Welded wire mesh boxes with stone fill
MMG – Carrington Weldgrip	MMG Ltd	Welded wire mesh boxes with stone fill
Titan	ABG Ltd	Woven polypropylene fabric cases with soil fill
Concertainer Bastions	Weldmesh Land Reinforcement/Tinsley Wire (Wigan)	Geotextile-lined mesh boxes generally filled with soil
Gabion mattresses		
Reno mattresses	Maccaferri Ltd	Double twisted wire mesh mattresses with stone fill
Weldmesh	Weldmesh Land Reinforcement/Tinsley Wire (Wigan)	Welded wire mesh mattresses with stone fill
MMG – Carrington Weldgrip	MMG Ltd	Welded wire mesh mattresses with stone fill
Triton Marine Mattresses	Tensar/Netlon	Polyethylene grid mattresses with stone fill
Sacks/Rolls		
Maccaferri	Maccaferri Ltd	Cylindrical gabions made of double twisted wire mesh and filled with stone
Bestmann Rock Rolls	MMG Ltd	Rolls of netting filled with rock
Bestmann Fibre Rolls	MMG Ltd	Coir rolls filled with soil and usually pre-planted
Greenfix Bio-rolls	Phi Group	Coir rolls with planting pockets
Concrete blocks		
Loose blocks		
Armorloc	MMG Ltd	Cellular dovetail-shaped blocks; interlocking
Ankalok	Ruthin Precast Concrete	Double anchor head shaped blocks; interlocking
Dytap	Ruthin Precast Concrete	Rock faced or plain interlocking blocks (also articulated)
Dycel	Ruthin Precast Concrete	Cellular or solid interlocking blocks (also articulated)
Basalton	Ruthin Precast Concrete	Loose, solid columns (also with a lava rock face)
Tri-lock	Grass Concrete Ltd	System of pairs of 'lock' and 'key' blocks
Channel-lock	Erosion Prevention Products, LLC, USA	Cellular interlocking blocks (also articulated)
Grassguard 160 & 180	Marshalls	Open grid paving blocks

Table A1.2 (continued) Proprietary revetment systems

Proprietary name	Manufacturer/supplier*	Brief description
Grasscel	Ruthin Precast Concrete	Loose, cellular paving blocks
Grassblock	Grass Concrete Ltd	Loose cellular paving blocks
Cabled blocks (or block mats)		
Armorflex	MMG Ltd	Cellular, interlocking blocks; also solid blocks
Petraflex	Grass Concrete Ltd	Stack-bonded cellular blocks; also solid blocks
Dycel	Ruthin Precast Concrete	Cellular or solid interlocking blocks (also loose)
Dytap	Ruthin Precast Concrete	Rock faced or plain interlocking blocks (also loose)
Channel-lock	Erosion Prevention Products, LLC, USA	Cellular interlocking blocks (also loose)
Other		
Flexible forms		
Fabriform	Proserve Limited	Woven mattresses filled with micro-concrete
Intrucell	Intrucell	Textile mattresses filled with pumped concrete
Concrete		
Grasscrete	Grass Concrete Ltd	In-situ cellular slabs
Soil reinforcement systems/ geomats		
Armater	MMG Ltd	Honeycomb structure in nonwoven polyester
Erosaweb GW	ABG Ltd	Honeycomb structure in polypropylene
MacMat	Maccaferri	Three-dimensional entangled polypropylene filaments (integrated with double twisted hexagonal mesh in Macmat-R)
Terramesh	Maccaferri	Double twisted hexagonal mesh with a front face consisting of stone-filled gabions (soil-filled in the Reinforced Green Terramesh)
Fortrac	MMG Ltd	Reinforcing geogrid made of high modulus polyester yarns with additional PVC layer
Textomur	Comtec (UK) Ltd	Reinforcing geotextile with steel mesh formwork covered by vegetation geotextile

Table A1.2 (continued) Proprietary revetment systems

Proprietary name	Manufacturer/supplier*	Brief description
Enkamat	MMG Ltd	Three-dimensional random polyamide matting (also filled with stone-bound bitumen and turfed-Enkazon)
Tensar Mat	Tensar/Netlon	Three-dimensional mat with a flat base layer and cuspated surface made of polyethylene; also Tensar Turf Mat and Macadamat
Geoweb	Cooper Clarke Group PLC	Cellular structure in high density polyethylene and ultrasonically welded joints
Erosamat Type 3	ABG Ltd	Random polyethylene filaments forming a thin mat

*For address see Table A1.3

Table A1.3. Addresses of manufacturers/suppliers of proprietary revetment systems

Address	Telephone	Facsimile
ABG Ltd Unit 01 Meltham Mills Knowle Lane Meltham West Yorkshire HD7 3DS UK	(0)1484 852096	(0)1484 851562
Comtec (UK) Ltd Bells Yew Green Tunbridge Wells Kent TN3 9BQ UK	(0)1892 750664	(0)1892 750660
Cooper Clarke Group Plc Bloomfield Road Farnworth Bolton BL4 9LP UK	(0)1204 862222	(0)1204 793856
Erosion Prevention Products, LLC 1328 NW Freeway Suite F244 Houston Texas 77040 USA	+713 729 9595	+713 728 3360
Grass Concrete Ltd Walker House 22 Bond Street Wakefield West Yorkshire WF1 2QP UK	(0)1924 375997 374818	(0)1924 290289
Intrucell 36 Benslow Rise Hitchin Hertfordshire SG4 9QY UK	(0)1462 454169	(0)1462 454169
Maccaferri Ltd 4B The Quorum Oxford Business Park Garsington Road Oxford OX4 2JY UK	(0)1865 770555	(0)1865 774550
Marshalls Southowram Halifax West Yorkshire HX3 9SY UK	(0)1422 306000	(0)1422 330185

Table A1.3. (continued) Addresses of manufacturers/suppliers of proprietary revetment systems

Address	Telephone	Facsimile
MMG Civil Engineering Systems Ltd St Germans King's Lynn Norfolk PE34 3ES UK	(0)1553 617791	(0)1553 617771
Phi Group Ltd Harcourt House Royal Crescent Cheltenham Gloucestershire GL50 3DA UK	(0)1242 510199	(0)1242 222569
Proserve Limited 80 Priory Road Kenilworth Warwickshire CV8 1LQ UK	(0)1926 512222	(0)1926 864569
Ruthin Precast Concrete Thornfalcon Works Henlade Taunton Somerset TA3 5DN UK	(0)1823 444606	(0)1823 444616
Tensar-Netlon Ltd New Wellington Street Blackburn BB2 4PJ UK	(0)1254 262431	(0)1254 694302
Weldmesh Land Reinforcement Tinsley Wire (Wigan) Woodhouse Lane Wigan Lancashire WN6 7NS UK	(0)1942 244071	(0)1942 824573

Category Box gabions
Product Maccaferri Gabions

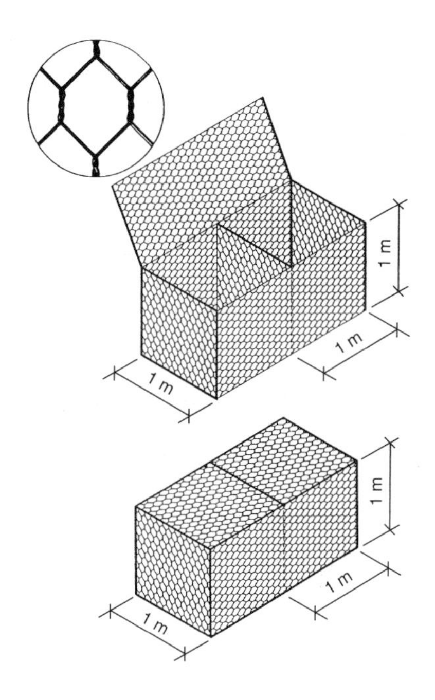

DESCRIPTION

Box-shaped containers usually filled with stone. The boxes are fabricated from a double twisted hexagonal mesh of galvanized wire; also available with 0·5 mm thick PVC coating providing continuous protection. Nominal mesh size: 80 mm.

Filling material consisting of round or quarried stone. Vegetation growth can be encouraged by adding top soil covers and by planting through the external face of the gabion above low water level. Pre-seeded and soil-filled lids made of MACMAT-R can also be used to assist vegetation growth.

CONSTRUCTION

Supplied folded, and usually assembled and filled on site. Heavy machinery not generally required. Can be prefilled and placed under water by means of lifting frames or pontoons.

PRODUCT DATA

Standard dimensions for box gabions are: 0·5 m to 1·0 m height, 1·0 m depth and 2 m to 4 m length with diaphragms 1 m apart. Other dimensions also available. Comprehensive information obtainable from manufacturer (including limiting flow velocities).

Category	Box gabions
Product	Weldmesh Gabions

DESCRIPTION

Box-shaped containers usually filled with stone. The box mesh can be made of steel wire or high density polyethylene (for contaminated soil or water). The steel mesh is electrically welded at every intersection and options include: hot dip galvanisation, bezinal coating and grey plastic coating bonded on galvanised wire. Nominal mesh sizes: 75 mm and 62 mm (square).

Filling material between 100 mm and 150 mm. Top soil cover can be used to promote vegetation growth. Also available in concertina form (Concertainer Bastions).

CONSTRUCTION

Supplied in collapsed form and filled on site. Heavy machinery not generally needed but crane installation of prefilled gabions may be a requirement in some cases. Gabion protection should be laid on a suitable geotextile membrane.

PRODUCT DATA

Standard sizes and specifications
Hot dip galvanised to BS729 or bezinal coated
Nominal dimensions — Twil Weldmesh specifications

Length × width × depth: m	Diaphragms	Capacity: m^3	Nominal mesh × wire diameter: mm
1·0 × 1·0 × 0·5	0	0·50	
1·0 × 1·0 × 1·0	0	1·00	
1·5 × 1·0 × 0·5	1	0·75	75 × 75 × 3·00
1·5 × 1·0 × 1·0	1	1·50	or
2·0 × 1·0 × 0·5	1	1·00	75 × 75 × 4·00
2·0 × 1·0 × 1·0	1	2·00	or
3·0 × 1·0 × 0·5	2	1·50	75 × 75 × 5·00
3·0 × 1·0 × 1·0	2	3·00	
4·0 × 1·0 × 0·5	3	2·00	
4·0 × 1·0 × 1·0	3	4·00	

Grey plastic coated bonded on wire galvanised to BS443
Nominal dimensions — Twil Weldmesh specifications

Length × width × depth: m	Diaphragms	Capacity: m^3	Nominal mesh × wire diameter: mm
1·0 × 1·0 × 0·5	0	0·50	75 × 75 × 2·70
1·0 × 1·0 × 1·0	0	1·00	galvanised core, then
1·5 × 1·0 × 0·5	1	0·75	coated respectively to
1·5 × 1·0 × 1·0	1	1·50	3·20 overall
2·0 × 1·0 × 0·5	1	1·00	
2·0 × 1·0 × 1·0	1	2·00	Heavier specifications
3·0 × 1·0 × 0·5	2	1·50	may be available upon
3·0 × 1·0 × 1·0	2	3·00	request
4·0 × 1·0 × 0·5	3	2·00	
4·0 × 1·0 × 1·0	3	4·00	

Category Box gabions
Product MMG – Carrington Weldgrip

DESCRIPTION

Box-shaped containers filled with stone. The mesh is electrically welded at each intersection; hot dip galvanised heavy duty steel wire can also be used if required. Nominal mesh size is 75 mm × 75 mm and wire gauge ranges from 3·1 mm to 5·0 mm.

The dimensions of the fill material should exceed the mesh size.

CONSTRUCTION

Supplied in fold flat form (base, four sides, dividers — if required — and lid). The components are assembled on site using either butt welded rings, ring clips or helicoils and then filled with stone. An overfill of 20 mm to 50 mm is recommended to allow for settlement.

PRODUCT DATA

Reference	Nominal dimensions: height × width × length	Dividers	Capacity: m^3
CW 1000	0·5 m × 1 m × 1 m	0	0·5
CW 1001	0·5 m × 1 m × 2 m	1	1·0
CW 1002	0·5 m × 1 m × 3 m	2	1·5
CW 1003	0·5 m × 1 m × 4 m	3	2·0
CW 1004	0·5 m × 1 m × 5 m	4	2·5
CW 1005	1 m × 1 m × 1 m	0	1·0
CW 1006	1 m × 1 m × 2 m	1	2·0
CW 1007	1 m × 1 m × 3 m	2	3·0
CW 1008	1 m × 1 m × 4 m	3	4·0

Category Box gabions
Product Titan

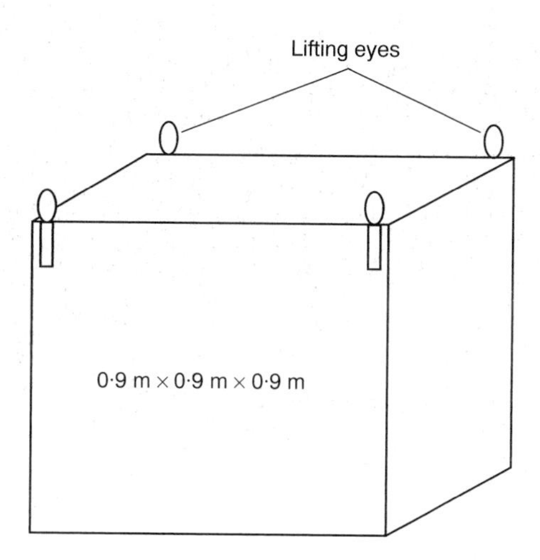

DESCRIPTION

Box-shaped containers of woven polypropylene filled with soil or cementitious fill. The tightly woven cage material allows the use of fine particle fills and the reuse of excavated materials. Nominal standard dimensions are 900 mm × 900 mm (base) by 900 mm (height) and weight of typically 2 t. Boxes of approximately half size are also available if required (500 mm × 500 mm × 500 mm).

Can be used as a light-weight embankment fill and in conjunction with a dry joint block facing.

CONSTRUCTION

Can be filled either by hand or machine using a supporting frame. This can be carried out in situ or beforehand and the containers can then be craned into position using the four sewn-in lifting eyes that are provided in each unit.

PRODUCT DATA

Type	Base width mm	Base length mm	Height mm	Volume m^3	Filled weight (typical) t
999	900	900	900	0·9	2·00

Category Box gabions/soil reinforcement systems
Product Concertainer Bastions

DESCRIPTION

Geotextile-lined mesh boxes filled with soil; extra galvanised mesh compartments can be left unlined at the front and filled with rock to obtain a stone-wall effect. A wide range of other fill materials can be used (e.g. sand, ballast, concrete, etc.) and concrete can also be sprayed on the front face to increase resistance to abrasion.

Particularly suitable for raising embankment levels in emergency situations.

CONSTRUCTION

Supplied in collapsed form and assembled by hand.

PRODUCT DATA

Manufactured from 'Weldmesh' wire mesh with a range of finishes: mild steel wire with zinc, bezinal or PVC coating.

Category Gabion mattresses
Product Reno Mattresses

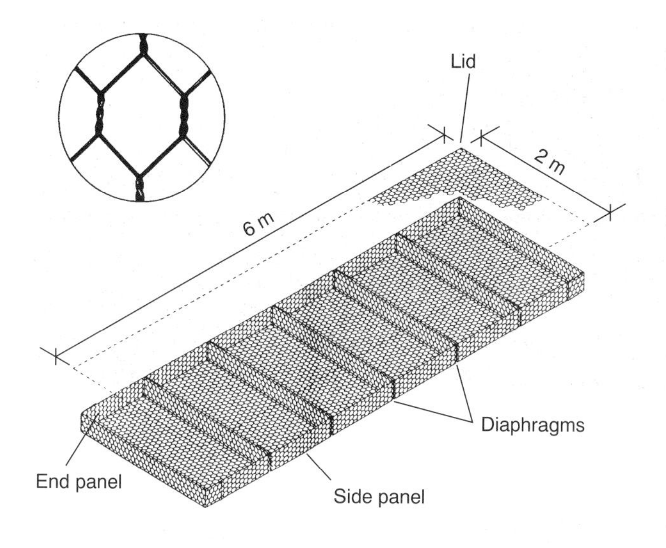

DESCRIPTION

Mattress-shaped containers usually filled with stone. A wide range of solutions is suggested by the manufacturer to promote vegetation cover and plant growth through the mattresses above mean water level. Sand asphalt mastic can also be used to grout the stone filling to achieve the required degree of permeability of the mattress and to increase resistance to current flow. The hexagonal mesh is made of double twisted galvanized wire; also available with 0·5 mm thick PVC coating. Lids of MACMAT-R filled with pre-seeded soil can be used to enhance vegetation. Maximum recommended bank slope is 1:1·5.

CONSTRUCTION

The mattresses are supplied folded and are usually assembled and filled on site. For underwater applications and in situations where mattresses are mastic grouted, they are prefabricated and lifted into position by cranes or launched with the aid of pontoons.

PRODUCT DATA

Comprehensive information obtainable from manufacturer (including limiting flow velocities) and various related products.

Type	Thickness: m	Filling stones	
		Stone size: mm	D_{50}
Reno mattresses	0·15–0·17	70–100	0·085
		70–150	0·110
	0·23–0·25	70–100	0·085
		70–150	0·120
	0·30	70–120	0·100
		100-150	0·125

Category Gabion mattresses
Product Weldmesh Mattresses

DESCRIPTION

Mattress-shaped containers usually filled with stone, thicknesses ranging between 150 mm and 300 mm. The box mesh can be made of steel wire or high density polyethylene (for contaminated soil and water). The steel mesh is electrically welded at every intersection and options include: hot dip galvanisation, bezinal coating and grey plastic coating bonded onto galvanised wire. Nominal mesh size: 75 mm $\times$ 75 mm.

Filling material between 100 mm and 150 mm. Top soil cover can be used to promote vegetation growth.

CONSTRUCTION

The mesh is supplied in collapsed form and mattresses are filled in situ. Prefilled mattresses can be lifted by means of a special crane and placed under water.

PRODUCT DATA

Standard sizes and specifications
Hot dip galvanised to BS729 or bezinal coated
Nominal dimensions — Twil Weldmesh specifications

Length × width × depth: m	Diaphragms	Capacity: m^3	Nominal mesh × wire diameter: mm
6 × 2 × 0·150	5	1·80	
6 × 2 × 0·225	5	2·70	
6 × 2 × 0·300	5	3·60	
3 × 2 × 0·150	2	0·90	75 × 75 × 2·50
3 × 2 × 0·225	2	1·35	
3 × 2 × 0·300	2	1·80	

Grey plastic coated bonded on wire galvanised to BS443
Nominal dimensions — Twil Weldmesh specifications

Length × width × depth: m	Diaphragms	Capacity: m^3	Nominal mesh × wire diameter: mm
6 × 2 × 0·150	5	1·80	75 × 75 × 2·70
6 × 2 × 0·225	5	2·70	galvanised core, then
6 × 2 × 0·300	5	3·60	coated respectively to
3 × 2 × 0·150	2	0·90	3·20 overall
3 × 2 × 0·225	2	1·35	
3 × 2 × 0·300	2	1·80	

Category Gabion mattresses
Product MMG – Carrington Weldgrip

DESCRIPTION

Mattress-shaped containers filled with stone. The mesh is electrically welded at each intersection and can be hot dip galvanised if required. Nominal mesh size is 75 mm × 75 mm and wire gauge ranges from 3·1 mm to 5·0 mm.

The dimensions of the fill should exceed the mesh size.

CONSTRUCTION

Supplied folded, the components are assembled on site and normally placed in position prior to filling. The components are assembled using either butt welded rings, ring clips or helicoils.

The manufacturer recommends that units should be overfilled by 20 mm to 50 mm to allow for settlement.

PRODUCT DATA

Reference	Nominal dimensions: height × width × length	Dividers	Capacity: m^2
CW 1009	0·15 m × 2 m × 6 m	5	1·8
CW 1010	0·30 m × 2 m × 6 m	5	3·6

Category Gabion mattresses
Product Triton Marine Mattresses

DESCRIPTION

Mattress-shaped containers usually filled with stone and made of high strength Tensar geogrids (polyethylene grids). The mattresses have internal baffles to prevent fill movement. They are suitable for underwater application since they can be lifted and placed in position while retaining their structural integrity.

CONSTRUCTION

The mattresses can be filled on or off site and will generally be placed over a prepared surface featuring a geotextile membrane. When underwater placement is required, the mattresses are prefabricated and transported by barge to the site.

PRODUCT DATA

Comprehensive technical data and design information are obtainable from the manufacturer.

Category Sacks/rolls
Product Maccaferri

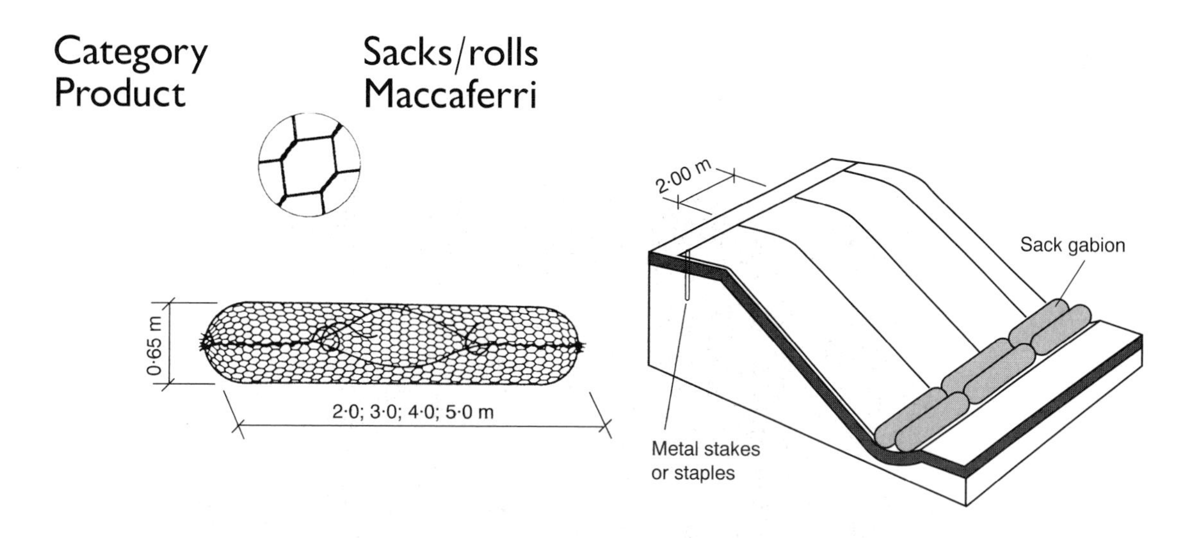

DESCRIPTION

Cylindrical gabions with sausage-like shape made of double twist hexagonal mesh and filled with stone. The mesh material is usually galvanised wire but PVC coating is also available.

Used mainly to fill scour holes due to their round shape and to provide toe embankment protection, particularly in emergency situations. They are also suitable for use in low strength soils.

CONSTRUCTION

Sack gabions are prefilled and placed in position by lifting equipment, in many cases under water, at the toe of embankments. They are not usually tied together but their round shape enables them to find an adequately stable position.

PRODUCT DATA

Refer to the manufacturer for detailed information.

Category Sacks/rolls
Product Bestmann Rock Rolls

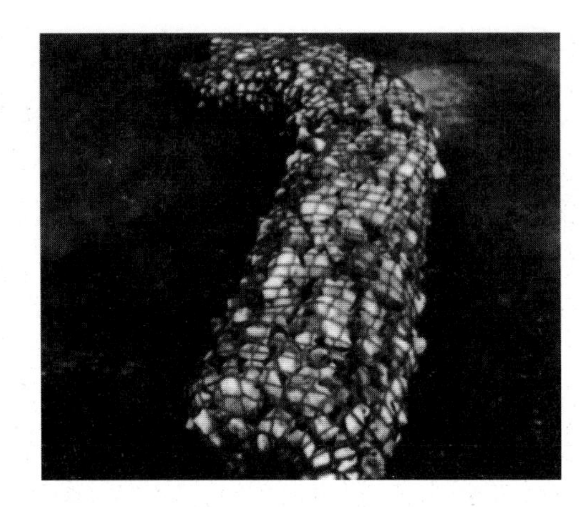

DESCRIPTION

Rolls of nylon netting filled with rock and used to provide erosion resistance at the toe of river banks. In many cases they are used as a foundation for Fibre Rolls. The flexibility of the rock rolls allows them to conform easily to the contours of river banks and to fill existing scour holes.

Preplanted rock rolls are also available, incorporating 5 cm × 5 cm root balls at the mean water level.

CONSTRUCTION

Normally supplied ready-made, rock rolls are lifted into position. When filling existing scour holes, no restraint is usually required but good fixing is important when rock rolls are used to protect the toe of a bank. Common methods include timber stakes and anchorage into the sloping bank.

PRODUCT DATA

Refer to the manufacturer for details.

Category Sacks/rolls
Product Bestmann Fibre Rolls

DESCRIPTION

Coir rolls developed to support plant growth along waterway banks. The rolls' fibre is biodegradable (life expectancy of a few years) but stability is expected to be maintained or increased by natural maturing of roots. The rolls are manufactured from coir fibre, which is homogeneously compressed, and contained by a polyethylene mesh of 50 mm openings. Root balls are usually included to promote quick establishment of vegetation. For environmental reasons native plants are chosen whenever possible.

Fibre rolls can be used as a single system for bank protection in smaller streams but can often be found in combination with rock rolls placed at the toe of a bank, or with sheet piling in situations of severe flow conditions. A wide range of composite solutions can be achieved.

CONSTRUCTION

The fibre rolls are supplied ready-made and usually pre-planted. They are lifted into position and require good restraint, particularly at the toe of banks. This can be achieved by placing vertical or sloping timber stakes about 1 m apart to contain the rolls on the flow side and by anchoring, also using stakes, into a sloping bank. A geotextile filter is usually recommended behind the rolls to prevent loss of backfill soil. When several layers of rolls are used it is advisable to tie the rolls together to ensure that the integrity of the embankment is maintained.

In cases where fibre rolls are used to 'soften' the hard appearance of sheet piling, adequate fixing to the piles' capping is very important.

PRODUCT DATA

Rolls of various diameters are available (e.g. 300 mm) and plant requirements can be accommodated by the supplier.

Category	Sacks/rolls
Product	Greenfix Bio-rolls

DESCRIPTION

Coir fibre rolls for protection of channel banks against erosion. The rolls are made of biodegradable fibre with planting pockets for establishment of plants along the waterline. In higher velocity flows the rolls can be enveloped in translucent polypropylene nets for added strength. The rolls are supplied pre-planted in most cases.

CONSTRUCTION

The fibre rolls are installed on channel slopes and fixed by means of stakes or steel pins so that the planting pockets face the bank. When in position, only the top 5–10 cm of the roll should be above the waterline. Rolls can also be stacked to form high slopes.

PRODUCT DATA

Rolls are supplied with 300 mm diameter and 3 m in length but other sizes can be made to order. Various types of plants can also be supplied to meet requirements.

Category Concrete blocks — loose
Product Armorloc

DESCRIPTION

Loose concrete blocks that interlock with six neighbouring units due to a double tapered dovetail shape; further interlocking is achieved in the vertical plane by blinding the blocks with gravel. The blocks have a cellular nature; when the cells are filled with topsoil, grass growth can occur above the waterline. Being hand laid, these blocks are recommended in situations where access is restricted and pre-assembled revetments are difficult to install.

CONSTRUCTION

Blocks are delivered palletised on crane off-load vehicles. Large debris and stones need to be removed from banks, and the banks should be graded to the required angle. A suitable geotextile should be placed under the blocks. Laid by hand, the blocks should be blinded with 2–20 mm gravel and/or grass-sown topsoil.

PRODUCT DATA

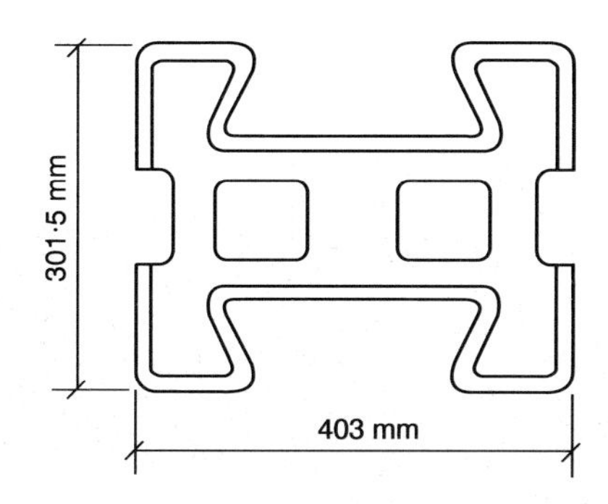

Armorloc data	
Gross area/block	0·09 m^2
Open area	20%
Mannings *n*	0·040

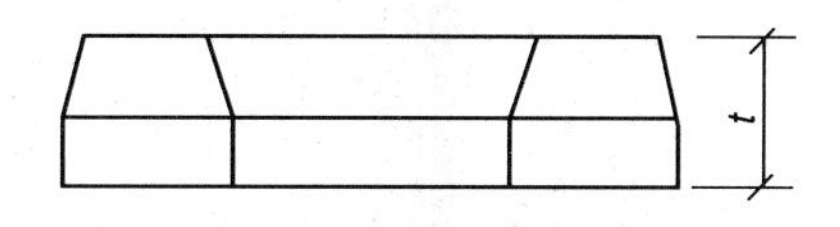

Concrete specification	
Density	2200 kg/m^3
Sulphate resistance	BS 5328: 1991 Class 2
Freeze thaw test	No visible effect

Type	Thickness *t*	Wt. of block	Wt. of block/m^2	No.of blocks/m^2
Armorloc	90 mm	12·8 kg	150 kg	11

Category Concrete blocks — loose
Product Ankalok

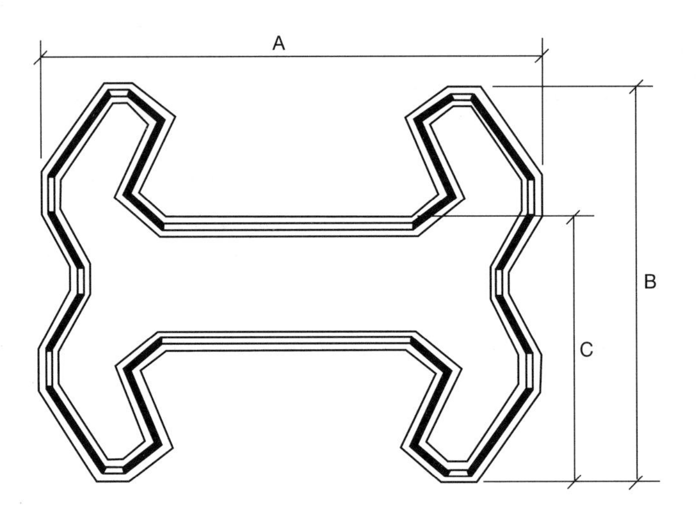

DESCRIPTION

Loose blocks with a double anchor head that interlock with six neighbouring units; additional frictional interlock is achieved by filling joints between blocks with gravel/soil. Seeded soil can be used to promote vegetation growth in the spaces between blocks above the water line. Below water level gravel can be used to add extra weight to the revetment. Being laid by hand, the blocks are particularly appropriate for sites with difficult access.

CONSTRUCTION

Blocks are supplied palletised. Areas to be protected need to be graded to required levels and cleared of obstructions. Where toe support is required but difficult to build, additional restraint can be provided by installing timber or steel pegs along the top of the revetment. Blocks are laid by hand and joints should be filled with appropriate material (e.g. topsoil or gravel).

PRODUCT DATA

Dimensions and weights

System reference	Thickness: mm	Module weight: kg	Installed weight: kg/m^2	Open area: %	Fill/m^2: m^3/m^2
Ankalok 90	90	12	150	25	0·025

Block size

A	B		C	
Actual length: mm	Actual width: mm	Effective length: mm	Effective width: mm	Coverage: No./m^2
380	300	400	200	12·5

Category Concrete blocks — loose
Product Dytap (also articulated — see Cabled blocks)

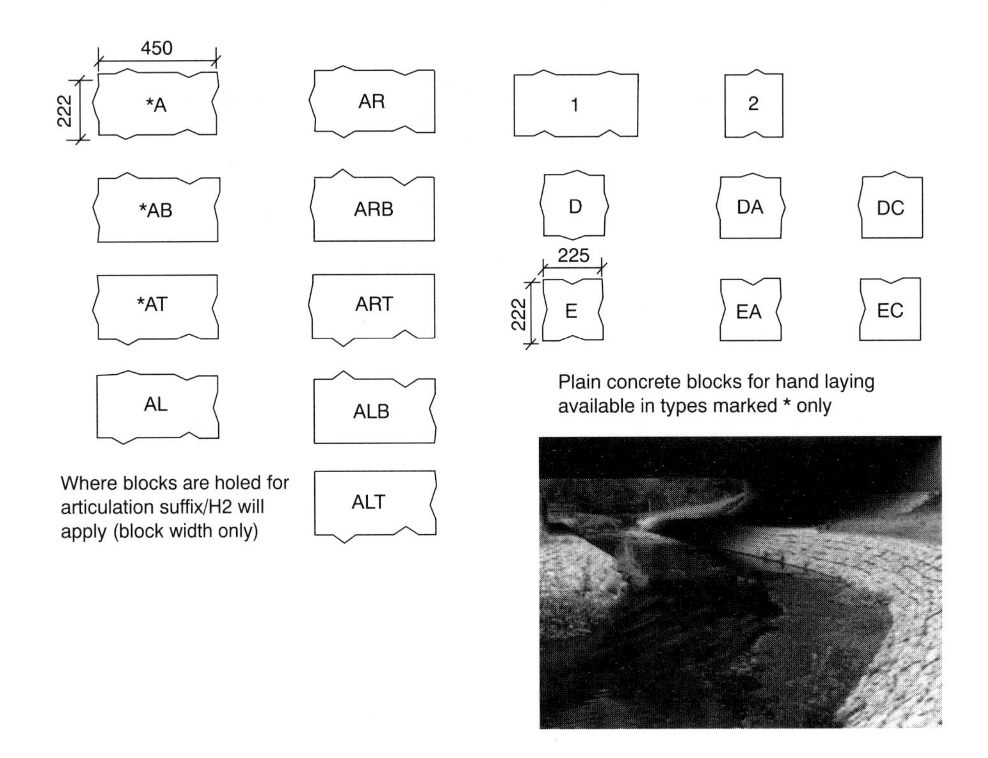

DESCRIPTION

Loose, solid interlocking blocks of jagged outline (also available in articulated mattresses — see Cabled blocks). The blocks can be plain or rock faced and provide an impermeable revetment. Hand laying should commence from toe to top on a surface prepared to the required levels and free of obstructions.

CONSTRUCTION

Blocks are laid dry jointed in stretcher bond pattern from toe to top of embankment. Areas to be protected need to be graded and soft points or holes should be filled. A suitably designed drainage layer of granular material and geotextile membrane will be required for large schemes in cohesive banks but may be omitted in smaller streams. Toe protection will generally be needed and may consist of a mass concrete toe.

PRODUCT DATA

Manufacture

Standard Dytap units are manufactured according to the following:

cement:	OPC to BS12: 1989
aggregates:	BS882: 1983
concrete mix design:	BS5328: 1991 Grade C35 Table 9
durability:	very severe exposure to Table 5 BS5328: 1991
sulphate resistance:	to Class 2, Table 7 BS5328: 1991
minimum cement content:	370 kg/m^3
maximum aggregate size:	10 mm
density:	2300 kg/m^3

Type reference	Facing	Nominal length: mm	Nominal width: mm	Nominal thickness: mm	Average unit: kg	Weight /m^2 kg	No. of blocks per/m^2
A	Masonry	450	222	75	21	210	10
A	Plain	450	222	75	17·5	175	10
A	Masonry	450	222	100	27	270	10
A	Plain	450	222	100	23	230	10
1	Masonry	450	222	150	41	410	10
1	Plain	450	222	150	34·5	345	10
1	Masonry	450	222	175	46	460	10
1	Plain	450	222	175	41	410	10
D,E	Plain	225	222	75	8·5	175	20
D,E	Plain	225	222	75	8·5	175	20
D,E	Masonry	225	222	100	14	280	20
D,E	Plain	225	222	100	12	240	20
2	Masonry	225	222	150	20	410	20
2	Plain	225	222	150	17	345	20
2	Masonry	225	222	175	23	460	20
2	Plain	225	222	175	20	410	20

Category Concrete blocks — loose
Product Dycel (also articulated — see Cabled blocks)

DESCRIPTION

Cellular or solid interlocking loose blocks with a jagged outline (also available in articulated mattresses — see Cabled blocks). Blocks are hand laid to a stretcher bond pattern. Seeded topsoil can be used to fill the open cells and joints and therefore encourage vegetation growth above the water-line; angular gravel is generally used below this level.

CONSTRUCTION

Blocks are laid dry jointed in stretcher bond pattern commencing from toe. Area to be protected needs to be prepared to required slope and levels. In most situations, a suitably designed drainage layer is recommended under the block revetment; this can be carried out by means of a geotextile membrane directly under the blocks or by granular filter layers. Toe protection will generally be required and may consist of a mass concrete toe.

PRODUCT DATA

Manufacture

Standard Dycel units are manufactured according to the following:

cement:	OPC BS12: 1989
aggregates:	BS882: 1983
concrete mix design:	BS5328: 1991 Grade C50 Table 9
durability:	very severe exposure to Table 5 BS5328: 1991

sulphate resistance:	to Class 2, Table 7 BS5328: 1991
minimum cement content:	370 kg/m^3
maximum aggregate size:	10 mm
density:	2300 kg/m^3 — when tested by the RILEM method with respect to exposure to repeated freeze/thaw, pods cut from actual Dycel units show no loss of weight

Dycel 100, 125 and 150

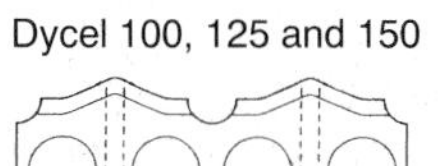

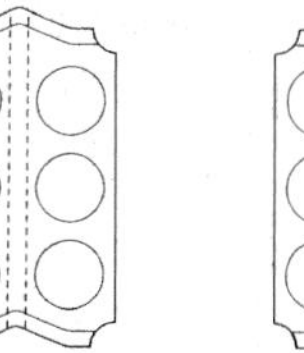

Dycel 101 and 151

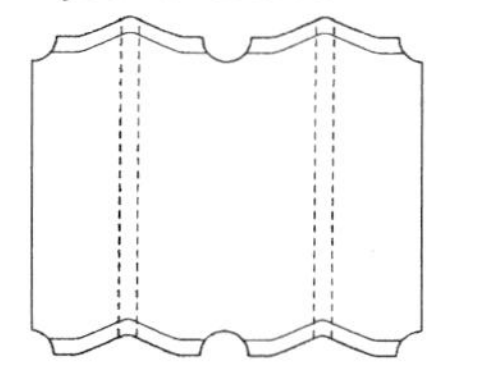

Dycel 220

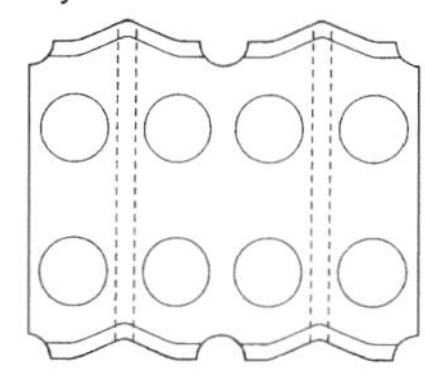

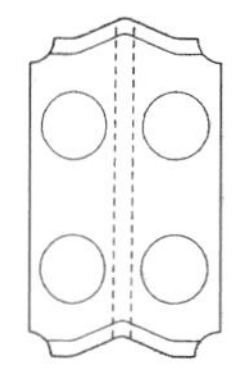

Dimensions and weights

System reference	Thickness: mm	Module weight: kg	Installed weight: kg/m^2	Open area: %	Fill/m^2: m^3/m^2
Dycel 100	100	30	155	30	0·030
Dycel 101	100	42	215	Solid	–
Dycel 125	125	37	190	30	0·0375
Dycel 150	150	45	230	30	0·045
Dycel 151	150	62	315	Solid	–
Dycel 220	220	80	410	16	0·035

Block type references

Block reference	Length: mm	Width: mm	Coverage: No./m^2
Type 1 (full block)	480	400	5·1
Type 2 (full block)	240	400	n/a

Category Concrete blocks — loose
Product Basalton and Basalton Eco

DESCRIPTION

Loose concrete columns of approximately hexagonal shape in plan. Placed in a closed polygon bond, they are tapered slightly upwards, which enables fast detection of subsidence since the columns will tend to tilt locally.

Colouring can be added or washed out from the heads of the columns to match surrounding work.

Basalton ECO columns have a top layer of coarse lava rock, 50 mm thick. This layer is exposed in the manufacturing process to trap silt and create a natural environment for plants and animal organisms.

CONSTRUCTION

The columns are laid with a specially designed machine clamp. The construction of an embankment revetment starts from the toe, abutting a row of stakes or toe retainer. On average, in each operation, the laying machine places sets of columns covering 1 m^2 area. Gravel is then placed in the interstices of the columns.

PRODUCT DATA

Column height	Density 2·3	
	t/m^2	m^2/t
15	0·312	3·21
20	0·416	2·40
25	0·521	1·92
30	0·628	1·59
35	0·710	1·41
40	0·811	1·23

Slight deviation in dimension and weight possible

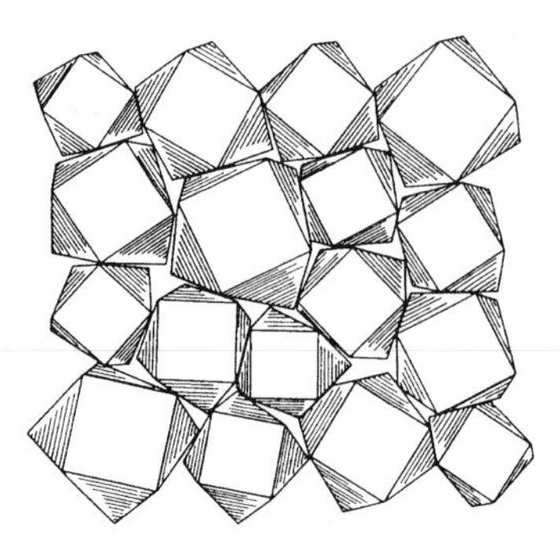

Category Concrete blocks — loose
Product Tri-lock

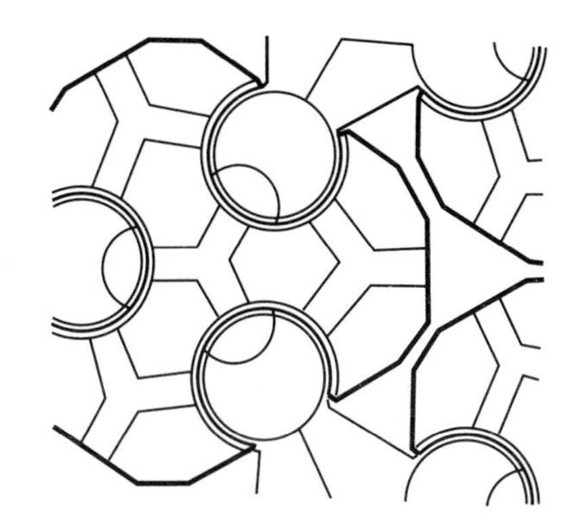

DESCRIPTION

System of pairs of loose concrete blocks formed by a 'lock block' and a 'key block'. Each block interlocks with three other units. Smooth land contours and changes in direction are achieved due to a bevel at the interlock and to the triangular nature of the system.

The blocks can be factory or in-situ made. They are placed by hand, usually over a carefully selected geotextile fabric which is normally supplied with the revetment blocks. Seeded topsoil can be used to fill the joints to encourage vegetation growth.

CONSTRUCTION

The system should be laid on a prepared surface, free of obstructions; slight variations in contour may be tolerated (not more than ± 40 mm in 6 m). A geotextile fabric is considered necessary in nearly all applications to ensure good drainage conditions.

When vegetation is established, maintenance is possible with conventional grass cutting equipment due to the evenness of the system.

PRODUCT DATA

Tri-lock blocks	Type 4	Type 5	Type 6
Height	90 mm	100 mm	150 mm
Weight/unit area	155 kg/m^2	171 kg/m^2	240 kg/m^2
Weight/pair of blocks (nominal)	23 kg	25 kg	35 kg
Concrete strength	> 40 N/mm^2	> 40 N/mm^2	> 40 N/mm^2
Open area (nominal, when laid)	30%	30%	30%

Category Concrete blocks — loose
Product Channel-lock
(also articulated — see Cabled blocks)

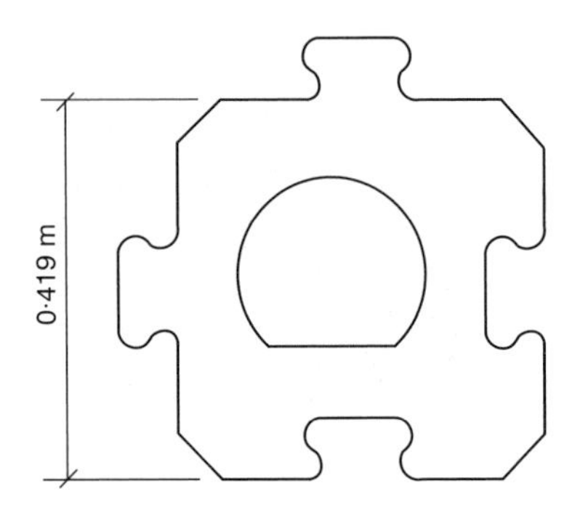

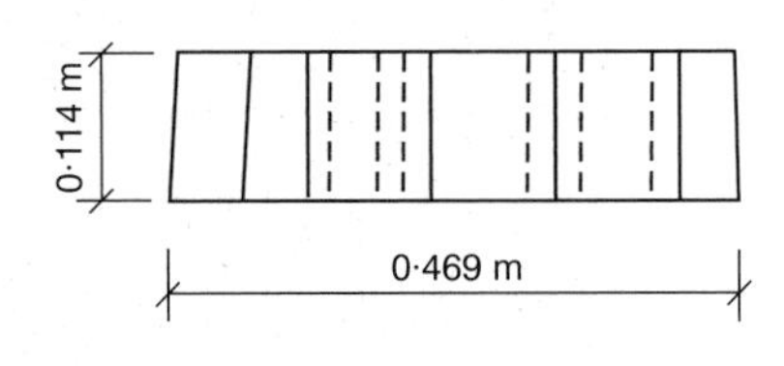

DESCRIPTION

Cellular interlocking blocks with an approximately octagonal shape. The interlocking effect is achieved mainly by connections on four of the sides. The blocks are wider at the bottom, tapering slightly towards the top face.

When filled with topsoil and seed, the voids in the blocks allow the growth of vegetation; the interlocking feature gives flexibility to follow irregular contours and to adjust to some change in the soil conditions.

CONSTRUCTION

Areas on which the blocks are to be placed should be free of voids and depressions and be levelled to the required grades. The manufacturer recommends the use of a filter fabric on top of which the blocks are laid. Installation of the blocks can be made by hand and should start from an upstream position to minimise the risk of damage caused by sudden flows.

At the perimeter, the revetment system should be turned and buried into the adjacent soil to a depth of at least 0·6 m. Grouting should be carried out at junctions with concrete walls, riprap, etc. to provide tight joints. The voids in the blocks are usually filled with topsoil, seed and fertilizer.

PRODUCT DATA

Blocks	450	550
Height: m	0·114	0·140
Weight per m^2	156	195
Surface area: m^2	0·165	0·165
Weight of block: kg	25·9	32·2
Concrete strength: N/mm^2	27·6	27·6
Open area (approx.)	20%	20%

Category — Concrete blocks — loose

Product — Grassguard 160 and 180

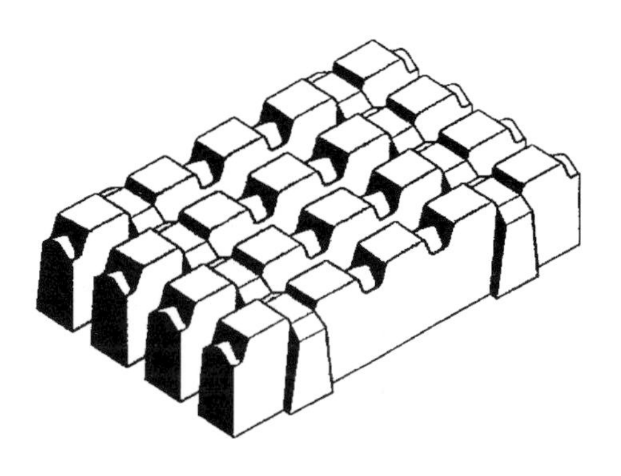

DESCRIPTION

Paving blocks forming a grid of concrete ribs and pockets of soil that allow vegetation growth and water percolation. The blocks are brown and rectangular in shape, and can be used for embankment stabilisation and berm reinforcement to create towpaths and vehicular access, among other uses. For stabilisation of embankments, the slopes must be set at the natural angle of repose of the soil.

CONSTRUCTION

The blocks should be laid onto a bedding/regulating layer of sharp sand or pea-shingle 25 mm thick; this is underlaid by a sub-base 80 mm to 150 mm thick (for Grassguard 160) and 150 mm (for Grassguard 180). On embankments the slope should not exceed 1:2 and an adequate toe restraint will be needed, for example by means of a toe beam. A geotextile fabric placed beneath the bedding layer may be beneficial to improve drainage.

In situations where flow velocities exceed 1 m/s the soil pockets should be filled with a 50/50 mixture of 6 mm gravel and well compacted seeded soil.

PRODUCT DATA

	Grassguard 160	Grassguard 180
Size: mm (nominal)	600 × 400 × 120	600 × 400 × 120
No. per m^2	4·17	4·17
Weight*: kg (each)	38	44
No. per pack	20	20
Weight*: kg (pack)	760	880
Colour	EARTH BROWN	EARTH BROWN
Approx. volume of fill material per m^2	0·05	0·04
No. of longitudinal ribs	4	5
No. of transverse ribs	2	2
% area of grass surface	75	75
% area of concrete base	76	82

*Pack sizes and weights may vary depending on factory of origin.

Category Concrete blocks — loose
Product Grasscel

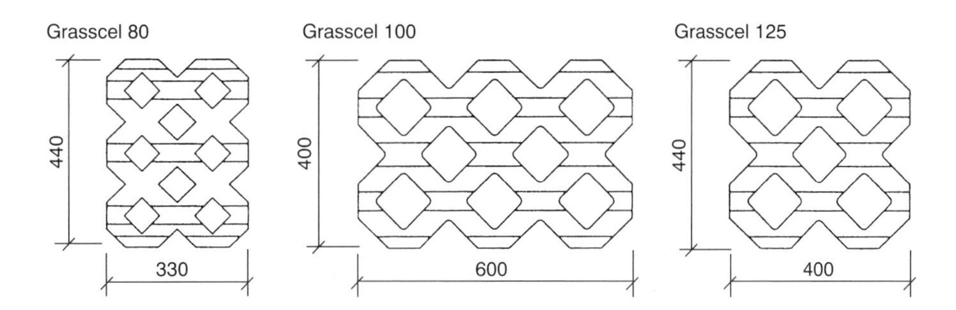

DESCRIPTION

Loose, cellular concrete blocks that enable extensive vegetation growth (75% of area). The system consists of blocks and raised platforms at regular intervals which form channels and provide a stable base for vehicle usage. These blocks are particularly suited for paving of channel berms that are required to provide access for maintenance or construction machinery.

CONSTRUCTION

Units are provided palletised and do not require special equipment for installation. Surfaces should be prepared to correct line and level, and be free from obstructions. The blocks should be installed with butt tight joints on a 20 mm layer of sharp sand overlaying a granular sub-base that has previously been well compacted. The open cells and joints should then be filled with topsoil and sown.

PRODUCT DATA

System	Grasscel 80	Grasscel 100	Grasscel 125
Plan size: mm	330 × 440	600 × 400	400 × 440
Thickness: mm	80	100	125
Open area at top: %	65	75	70
Unit weight: kg	17	29	30
Units/m^2: No.	6·89	4·167	5·68
Installed weight: kg/m^2	120	120	170
Fill required: m^3/m^2	0·03	0·045	0·050
Transverse: kN	10	11	22
Test BS7263			

Category Concrete blocks — loose
Product Grassblock

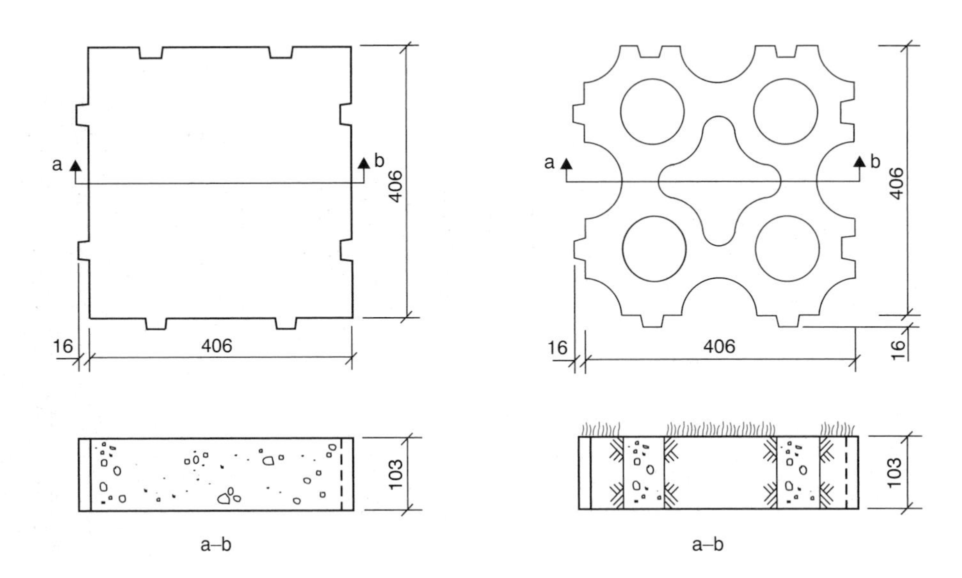

DESCRIPTION

Loose, cellular blocks with an open matrix that allows the establishment of a grass cover. The blocks have interlocking features and can be laid in two different patterns, one of which has less strength but bigger area of grass cover. Solid blocks are also available, as are brown-coloured blocks, if required. This system is suited for bank protection and for paving of berms subjected to light or moderate traffic loads.

CONSTRUCTION

The blocks are usually placed by hand in dry conditions on a trimmed slope blinded with a 20 mm thick layer of sharp sand. A nonwoven geotextile should be placed over this underlayer for all flow applications. After placing the blocks, the cells are then filled with seeded topsoil or filled with gravel below the waterline.

PRODUCT DATA

Block type	Dimensions: mm	Unit weight: kg	% open area	
			Surface	Base
Grassblock 103	406 × 406 × 103	22	43	34
Grassblock 103 solid	406 × 406 × 103	44	n/a	n/a
Grassblock 125	406 × 406 × 125	29	–	–

Category Cabled blocks
Product Armorflex

DESCRIPTION

Cellular or solid concrete blocks linked longitudinally by galvanised wire cables or polyester ropes. Interlocking stretcher bond between blocks provides linkage in the other direction but cabling can also be used in severe situations. A double taper on the sides of the blocks allows articulation of the joints and additional frictional resistance is obtained when these are blinded with gravel. The system provides a flexible mat suitable for river bank and bed protection that can also withstand wave attack.

The open structure of the cellular blocks helps to prevent build-up of back pressure behind the revetment and allows vegetation growth through the cells above mean water level.

CONSTRUCTION

Good preparation and compaction of the sub-grade is necessary prior to placement of a geotextile fabric and of the block mats. These are delivered assembled and a lifting beam is used to place them in position; for slopes of 1:2 or steeper the mats may slide down on the geotextile until they reach the required position. The methods of anchoring the mats will depend on the situation. They can range from temporary anchorage with wooden pegs along the top of the mat when vegetation is expected to develop and bind the mat to the sub-grade, to permanent anchorage using concrete beams, and ground anchors with non-corroding cables.

Preseeded top soil can be used to blind the mat and encourage plant growth.

PRODUCT DATA

Cellular block mats

Armorflex 140

L × B × H:	340 × 400 × 85 mm
Block weight:	17·7 kg each
Unit weight:	140 kg/m^2
Open area (min.):	20%

Armorflex 140S

L × B × H:	340 × 400 × 90 mm
Block weight:	19 kg each
Unit weight:	150 kg/m^2
Open area (min.):	20%

Armorflex 180

L × B × H:	340 × 300 × 120 mm
Block weight:	17·3 kg each
Unit weight:	180 kg/m^2
Open area (min.):	20%

Armorflex 220

L × B × H:	340 × 300 × 150 mm
Block weight:	21·9 kg each
Unit weight:	225 kg/m^2
Open area (min.):	20%

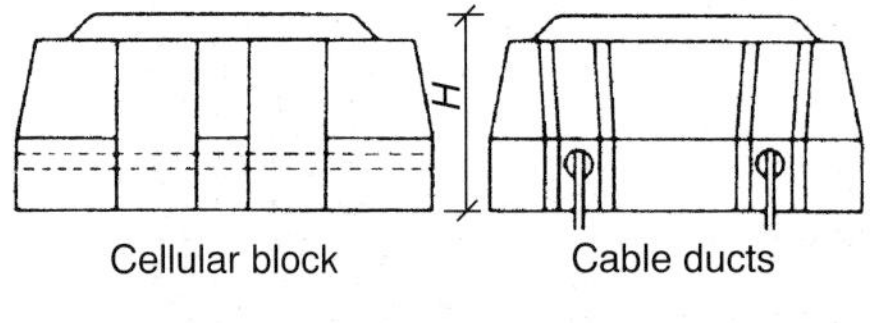

Cellular block   Cable ducts

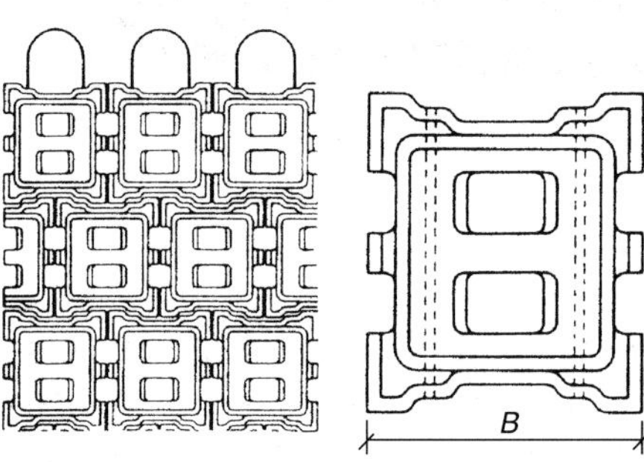

Solid block mats

Armorflex 165

L × B × H:	340 × 400 × 85 mm
Block weight:	20 kg each
Unit weight:	160 kg/m^2
Open area (min.):	6%

Armorflex 215

L × B × H:	340 × 300 × 120 mm
Block weight:	21 kg each
Unit weight:	220 kg/m^2
Open area (min.):	6%

Armoroc 215

L × B × H:	340 × 300 × 135 mm
Block weight:	22·85 kg each
Unit weight:	240 kg/m^2
Open area (min.):	6%

Armorflex 305

L × B × H:	340 × 300 × 150 mm
Block weight:	26·7 kg each
Unit weight:	285 kg/m^2
Open area (min.):	6%

Armoroc 305

L × B × H:	340 × 300 × 165 mm
Block weight:	28·65 kg each
Unit weight:	300 kg/m^2
Open area (min.):	6%

Armorflex 435

L × B × H:	340 × 300 × 225
Block weight:	41 kg each
Unit weight:	435 kg/m^2
Open area (min.):	6%

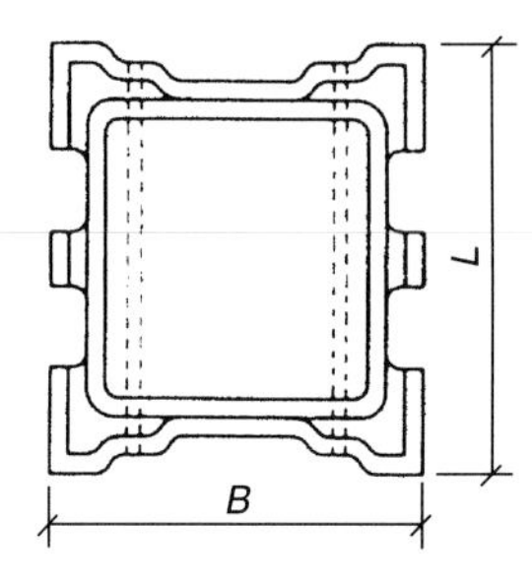

Category Cabled blocks
Product Petraflex

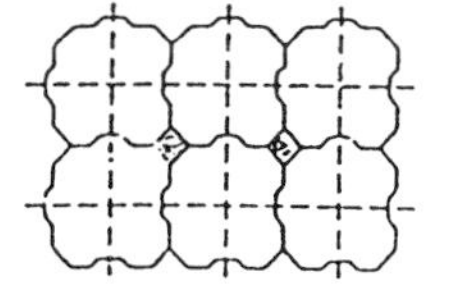

DESCRIPTION

Stack-bonded blocks (available as cellular, solid and natural stone faced) tied by cables in two orthogonal directions. The cables are manufactured in polyester and blocks of different types and thicknesses can be incorporated in the same panel to absorb wave energy and reduce wave run-up.

The mats can be installed under water with a geotextile fabric attached to the mats before placement.

CONSTRUCTION

Panels are delivered for installation usually in 2·44 m width by 6·1 m length. The area to be protected needs to be well prepared and adequate drainage layers under the mats (consisting of geotextile membrane(s) and filter layers of 20 mm clean aggregate) are considered essential to achieve the full potential of the revetment.

Installation is carried out by crane using single or double-end suspension. Adequate anchorage can be provided by fixing the mats at the top of the bank; ground anchors or trenches can be used to fix the panels at the toe.

PRODUCT DATA

Type	Unit coverage: mm × mm	Thickness: mm	Unit weight (nominal): kg	Superficial weight (nominal): kg/m^2	Open area: %	
					Top	Base
P3i	305 × 305	90	13	140	35	24
P4i	305 × 305	100	17	185	35	22
P5i	305 × 305	135	21	225	35	18
P9i	460 × 610	220	116	415	31	18
P4i solid	305 × 305	100	21	230	nil	nil
P5i solid	305 × 305	135	31	330	nil	nil
P9i solid	460 × 610	220	154	550	nil	nil

Category Cabled blocks
Product Dycel (also loose — see Loose blocks)

DESCRIPTION

Cellular or solid interlocking blocks with a jagged outline and linked in one direction with non-corrosive flexible tendons (also available as loose blocks — see Concrete blocks — loose). The block mats are particularly suited for underwater applications and where construction of an adequate toe support is not feasible.

The open structure of cellular blocks allows release of back-water pressure and when cells are filled with topsoil and seeded, plant growth is encouraged above mean water level. Solid blocks are used in very abrasive environments and near the toe of the banks to provide additional weight.

CONSTRUCTION

The surface to take the revetment needs to be prepared to required levels and be free from obstructions. In situations where geotextile fabric is recommended, overlaps should not be less than 300 mm. The mats are usually delivered factory-assembled; once the mats are laid, the cables can be fixed by ground anchors at the top of the bank or into in-situ concrete beams.

PRODUCT DATA

Manufacture

Standard Dycel units are manufactured according to the following:

cement:	OPC BS12: 1989
aggregates:	BS882: 1983
concrete mix design:	BS5328: 1991 Grade C50 Table 9
durability:	very severe exposure to Table 5 BS5328: 1991
sulphate resistance:	to Class 2, Table 7 BS5328: 1991
minimum cement content:	370 kg/m^3
maximum aggregate size:	10 mm
density:	2300 kg/m^3 — when tested by the RILEM method with respect to exposure to repeated freeze/thaw, pods cut from actual Dycel units show no loss of weight.

Dycel 100, 125 and 150

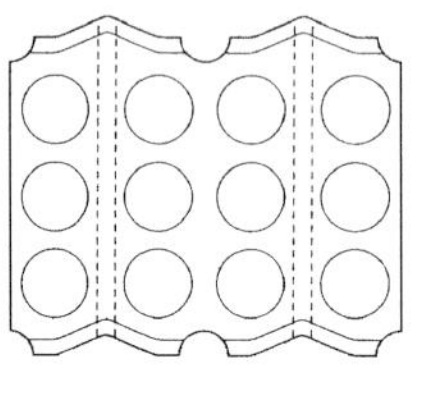

Dycel 101 and 151

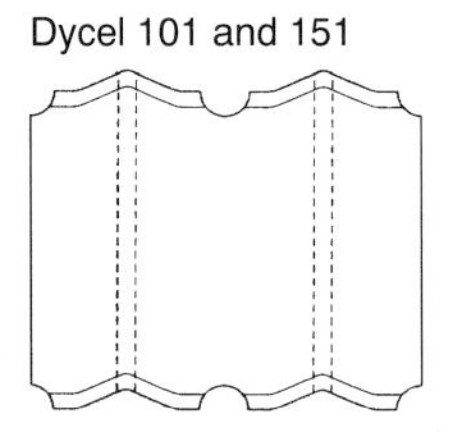

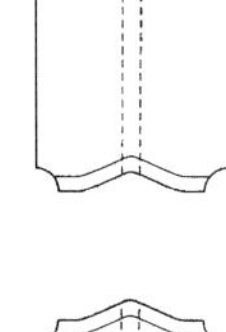

Dycel 220

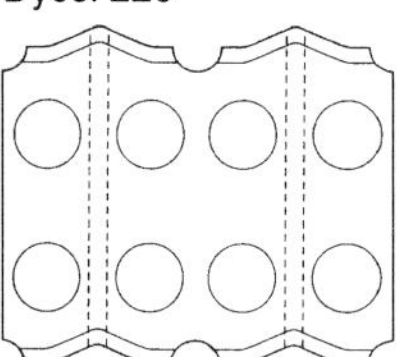

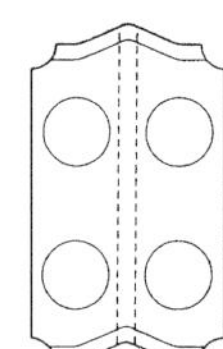

Dimensions and weights

System reference	Thickness: mm	Module weight: kg	Installed weight: kg/m^2	Open area: %	Fill/m^2: m^3/m^2
Dycel 100	100	30	155	30	0·030
Dycel 101	100	42	215	Solid	–
Dycel 125	125	37	190	30	0·0375
Dycel 150	150	45	230	30	0·045
Dycel 151	150	62	315	Soild	–
Dycel 220	220	80	410	16	0·035

Block type references

Block reference	Length: mm	Width: mm	Coverage: No./m^2
Type 1 (full block)	480	400	5·1
Type 2 (full block)	240	400	n/a

Category Cabled blocks
Product Dytap (also loose—see Loose blocks)

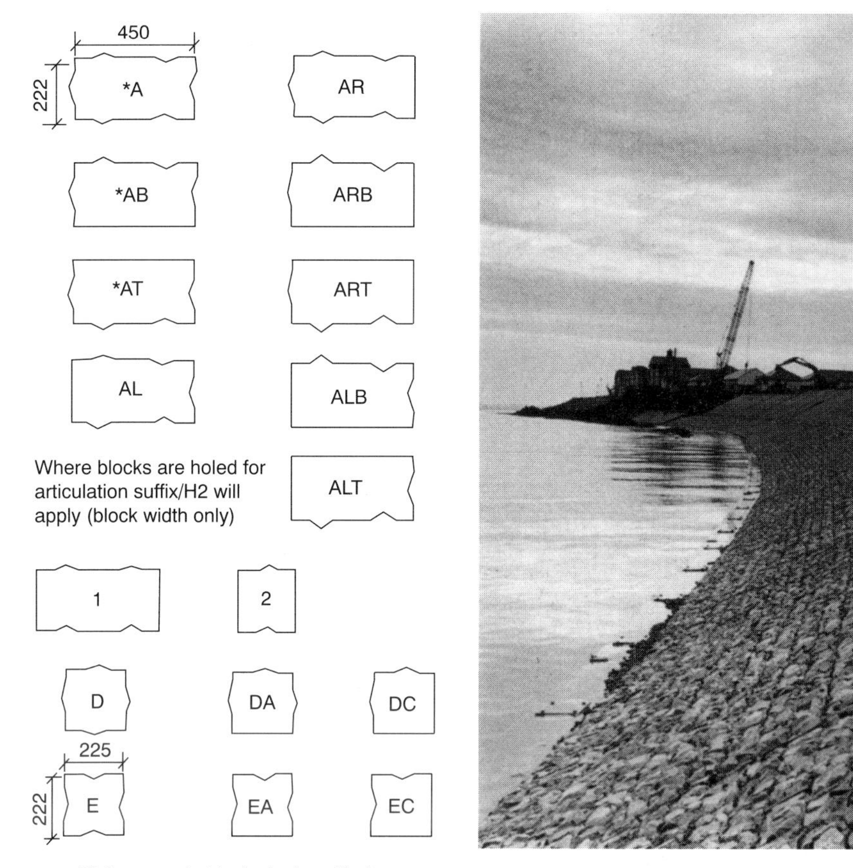

DESCRIPTION

Solid interlocking blocks with a jagged outline and linked through the smaller side to form articulated mats (also available as loose blocks — see Concrete blocks — loose). The linkage is provided by non-corrosive flexible tendons. The block mats are particularly suited where underwater placing is a requirement or where construction of adequate toe supports is not possible. The blocks can be plain or rock faced and provide an impermeable revetment.

CONSTRUCTION

The mats are usually delivered assembled but on-site fabrication is also possible. Areas to be protected need to be graded to correct levels and free from obstructions or holes. A suitably designed drainage layer should take into account the nature of the site soil and hydraulic loads.

The mats are normally supported by attaching the cables to a concrete anchor at the top of the bank; some form of toe protection is also recommended. The placing of panels into position can be done by single or double-end lifting, depending on panel weight and size.

PRODUCT DATA

Manufacture

Standard Dytap units are manufactured according to the following:

cement:	OPC to BS12: 1989
aggregates:	BS882: 1983
concrete mix design:	BS5328: 1991 Grade C35 Table 9
durability:	very severe exposure to Table 5 BS5328: 1991
sulphate resistance:	to Class 2, Table 7 BS5328: 1991
minimum cement content:	370 kg/m^3
maximum aggregate size:	10 mm
density:	2300 kg/m^3

Type reference	Facing	Nominal length: mm	Nominal width: mm	Nominal thickness: mm	Average unit: kg	Weight /m^2: kg	No. of blocks per/m^2
A	Masonry	450	222	75	21	210	10
A	Plain	450	222	75	17·5	175	10
A	Masonry	450	222	100	27	270	10
A	Plain	450	222	100	23	230	10
1	Masonry	450	222	150	41	410	10
1	Plain	450	222	150	34·5	345	10
1	Masonry	450	222	175	46	460	10
1	Plain	450	222	175	41	410	10
D,E	Plain	225	222	75	8·5	175	20
D,E	Plain	225	222	75	8·5	175	20
D,E	Masonry	225	222	100	14	280	20
D,E	Plain	225	222	100	12	240	20
2	Masonry	225	222	150	20	410	20
2	Plain	225	222	150	17	345	20
2	Masonry	225	222	175	23	460	20
2	Plain	225	222	175	20	410	20

Category Cabled blocks
Product Channel-lock (also loose — see Loose blocks)

DESCRIPTION

Mattresses formed by cellular interlocking blocks with an approximately octagonal shape and linked by polyester cables. The interlocking effect is achieved mainly by connectors on four of the sides. The blocks are wider at the bottom, tapering slightly towards the top face.

This system is suitable for underwater installation; block voids above the water-line can be filled with topsoil and seeded to promote vegetation growth.

CONSTRUCTION

Areas where the mattresses are to be placed should be free of voids and depressions and be levelled to the required grades. The manufacturer recommends the use of a filter fabric on top of which the mattresses are laid. Placement of mats should be done with the aid of a spreader bar and lifting crane. Adjacent mats are fixed together by connecting lateral cables and end cable loops. Grouting should be carried out to cover any gaps bigger than 50 mm. The voids in the blocks are usually filled with topsoil, seed and fertilizer.

PRODUCT DATA

Blocks	450	550
Height: m	0·114	0·140
Weight per m^2	156	195
Surface area: m^2	0·165	0·165
Weight of block: kg	25·9	32·2
Concrete strength: N/mm^2	27·6	27·6
Open area approx.	20%	20%
Mattresses (with cables)		
Standard width: m	2·44	2·44
Standard length: m	12·2	12·2

Category Flexible forms
Product Fabriform

DESCRIPTION

Woven mattresses made of nylon, polyester, polypropylene or other high strength geotextile and usually filled with a fluid concrete (micro-concrete) typically with a water/cement ratio between 0·65 and 0·70, and 2:1 sand/cement ratio. Although most common in white, the mattresses can also be fabricated in other colours to blend more effectively with the environment.

In its more basic form, the system produces rigid concrete slab protection and is therefore suitable for use over compact, stable banks. Where settlement of the ground is expected, the RB Fabriform Mattress is a more appropriate choice. In this flexible system, the mortar is injected into individual fabric pockets which are connected to adjacent blocks with strong nylon rope.

Fabriform is particularly suitable for underwater applications because concreting can be conducted in flowing water. Applications include large river bank and bed reaches, the repair of jetties, bridge piers, ship dockings, culvert protection and pipeline foundations. Reinforcement can be introduced inside the mattresses for additional strength.

CONSTRUCTION

Since the revetment takes the shape it was designed to fit, a perfect preparation of the base is not strictly necessary but it is important to secure the edges of the mattresses well, particularly at the toe of banks. The mattresses can be filled above or below water and be joined to adjacent units by special zips or secured with bolts, straps and ties.

PRODUCT DATA

Refer to supplier/manufacturer for technical data and design advice.

Category	Flexible forms
Product	Intrucell

DESCRIPTION

Mattresses made of two or more layers of high strength textile filled with pumped concrete. They are available in various thicknesses ranging from 75 to 1200 mm. The greater thicknesses are more appropriate for offshore and coastal applications. Mattresses 75 to 100 mm thick are generally selected for river bank protection; for bed protection and particularly for underwater applications, thicker mattresses are used. In these situations the product range Ovolo, which is supplied in 300 to 1200 mm thickness, may be considered. When filled, this product is formed by diamond-shaped compartments with flexible joints at their perimeter. This articulation feature is intended to allow some ground settlement without jeopardising the integrity of the revetment. Apertures in the revetment can be introduced for the relief of hydrostatic pressure.

CONSTRUCTION

The ability of the revetment to follow minor variations of the ground profile implies that ground preparation may be relatively basic. The textile mattresses are laid on the area to be protected and filled with pumped concrete. This can be done above or below water. The outer base layers of the mattresses can be designed to act as geotextiles and therefore prevent the loss of solids from the underlying soil.

PRODUCT DATA

Refer to supplier/manufacturer for technical data and design advice.

Category Concrete — cellular slabs
Product Grasscrete

DESCRIPTION

Cast in situ reinforced cellular slabs made of concrete. This system can be used for paving and also for slope protection in river and channel banks. The large open area of the cells facilitates the establishment of a good grass cover above the water-line. Below mean water level the cells should be filled with gravel.

CONSTRUCTION

The revetment should be built on a well consolidated sub-base with a 10–20 mm thick regulating sand layer. The formers and reinforcement mesh are then placed in position and concrete is poured over. During this process no flow should be allowed for about 48 hours. Exposed tops of formers are burnt out and the openings are filled with topsoil. This soil is then sown with a suitable grass mix to create a vegetation cover. Expansion joints are recommended at 10 m intervals.

PRODUCT DATA

Grasscrete® in situ reinforced concrete cellular paving			Depth: mm	Weight/m^2	% void	
					Surface	Base
Type	GC1	(A193 reinforcement)	100	n/a	47	53
	GC2	(A252 reinforcement)	150	n/a	42	56

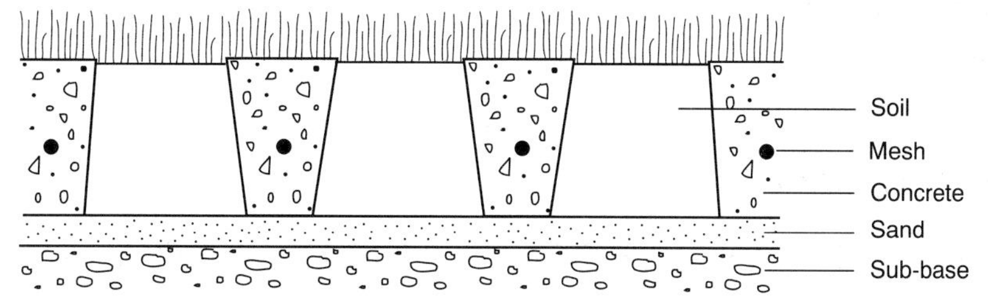

Grasscrete GC2 150 mm deep

Category Soil reinforcement systems/geomats
Product Armater

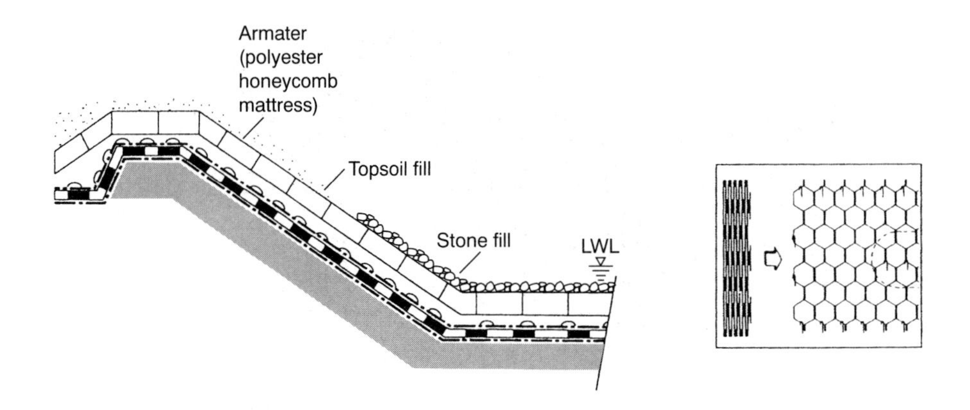

DESCRIPTION

Soil reinforcement system with honeycomb structure made by alternate linking strips of polyester material. The hexagonal cells can be filled with earth, gravel or concrete. Seeded soil can be used or small shrubs can be planted to help to establish a vegetation cover. Armater made of woven jute fabric is available where a biodegradable structure is required; Armater with a reinforcing fabric on the lower face can also be used to protect underlying channel linings.

CONSTRUCTION

Slopes need to be levelled and compacted prior to spreading of the panels, which are 150 m^2 in area and supplied rolled up. The panels are fixed at the top of the slope by means of an anchorage shelf (200 mm deep by 500 mm minimum width). Steel pins and buckles are used to fix the panels and temporary toe fixing is also required during filling. On steep slopes, intermediate pinning of 1 pin/m^2 is recommended. Filling can be carried out manual or mechanically and should preferably start by filling cells on diagonals along the slope.

PRODUCT DATA

AV 20-25/10 EC	Side lengths = 200 mm and 250 mm
	Height = 100 mm
AV 20-20/10 EC	Side length = 200 mm
(used for steeper slopes)	Height = 100 mm

Other cell sizes are available to special order.

Category Soil reinforcement systems/geomats
Product Erosaweb GW

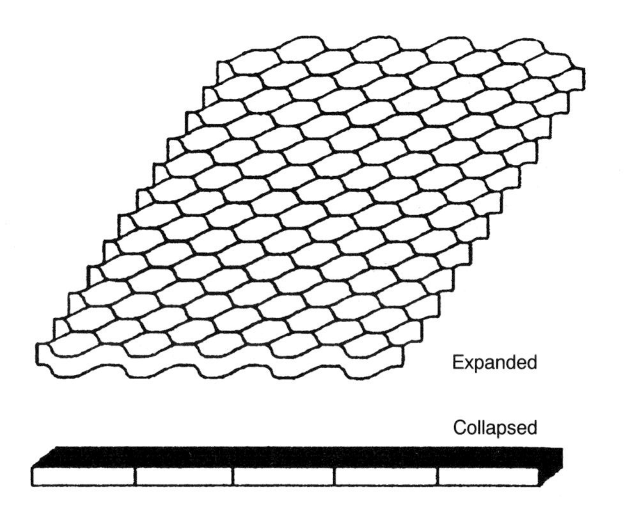

DESCRIPTION

Honeycomb structure of interconnecting polymer strips forming cells that can be filled with soil, stone or concrete. The strips are manufactured from high density polypropylene and are sewn together. When filled with topsoil and seeded above mean water level a vegetation cover can be established. Structures of various heights are available (e.g. 100 mm, 300 mm) as well as several cell sizes. A related product made from geotextile strips (Erosaweb AR) is also supplied by the manufacturer and is suitable for shallow slopes in above water applications.

CONSTRUCTION

Supplied folded flat and expanded on site, this system should be placed over a smooth profile and a nonwoven geotextile (manufacturer's recommendation). The connection between mattresses is achieved by simple interlocking of the jagged edges. Erosaweb 100 mm, 150 mm and 200 mm should be fixed at the top of the bank by means of an anchor trench 300 mm wide × 450 mm deep; Erosaweb 300 mm, 400 mm and 500 mm should be taken above the top of the slope and securely pinned. Steel pins at specified intervals should also be used to fix the mats along the slope. Filling should commence from the bottom of the slope and be carried out by machine. Below water level, crushed stone fill is recommended.

PRODUCT DATA

Depth	100/150/200	500	Special
Material	Polypropylene	Polypropylene	
Web thickness	1·5 mm	1·5 mm	Properties defined for specific schemes
Colour	Black	Black	
Effective cell diameter	275 mm	475 mm	
Cell area	0·03 m^2	0·18 m^2	
Temperature range	−30° to 60°C	−30° to 60°C	
Material tensile strength	45 kN/m	45 kN/m	
Seam tensile strength	1·5/2·2/3·0 kN	8·0 kN	
UV stability	Excellent	Excellent	
Standard pin length	450–600 mm	600–1200 mm	
Life expectancy (incl. joints)	120 years	120 years	
Max. slope angle	70°	70°	

Category Soil reinforcement systems/geomats
Product MacMat/MacMat-R

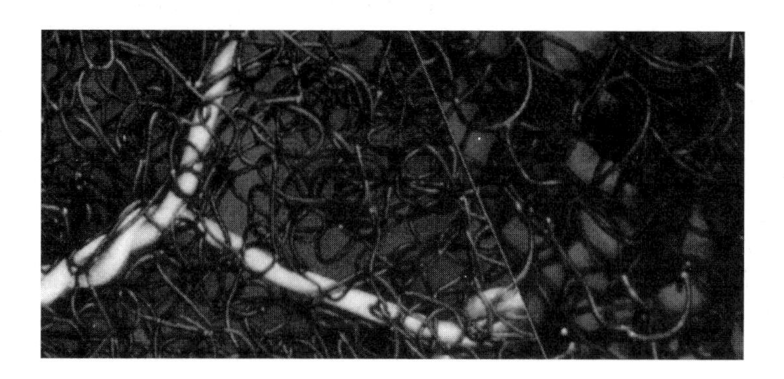

DESCRIPTION

Three-dimensional mat made of entangled polypropylene filaments, heat bonded at contact points. The mat can be placed directly over a pre-seeded surface or alternatively be filled with pre-seeded topsoil to speed up the establishment of vegetation cover. The erosion resistance of the system is increased by plant root growth through the mat.

MacMat-R is a related product which combines the three-dimensional geomat with a double twisted hexagonal mesh. This is used in more severe situations such as steep slopes and also as lids to Reno mattresses or gabions.

CONSTRUCTION

The mats are placed on top of well prepared bases and fixed at top and bottom of slopes with pins or staples and/or by anchor trenches. Edges should be pinned, typically at 1 m to 2 m centres. Mats can be allowed to fill naturally or be filled with seeded topsoil and overfilled by 10 mm to 20 mm. Hydroseeding or hydromulching are other alternatives. In areas frequently under water, gravel (2 mm to 5 mm) or gravel/soil mixtures may be advantageous.

PRODUCT DATA

	Item	Standard	Units	MacMat	MacMat-R	
					Zinc	Zinc + PVC
Polymer properties	Polymer type	–	–	Polypropylene		
	Density	ASTM D792	g/m^3	900		
	Melting point	ASTM D1505	°C	150		
	Colour	–	–	Black		
	UV resistance	ASTM D4355	–	Stabilised		
Roll size	Width	–	m	2	2	2
	Length	–	m	25	25	25
	Weight	–	kg	33	75	83
MacMat properties	Thickness	ASTM D1777	mm	15	10	10
	Unit weight	ASTM D3776	g/m^2	650	550	550
	Tensile strength	ASTM D4595	kN/m	> 1·6	n/a	n/a
	Upper void space	–	%	90	90	90
Mesh properties	Mesh type	UNI 8018	mm	n/a	60 × 80	60 × 80
	Wire core diameter	UNI 3598	mm	n/a	2·2	2·2
	Wire tensile strength	UNI3598	kg/mm^2	n/a	38–50	38–50
	Zinc coating	UNI 8018	g/m^2	n/a	240	240
	PVC outer diameter	UNI 3598	mm	n/a	n/a	3·2
	Mesh tensile strength	UNI 8018	kN/m	n/a	28	28

Note: other MacMat thicknesses and/or Maccaferri mesh configurations available on request

Category Soil reinforcement systems/geomats
Product Terramesh

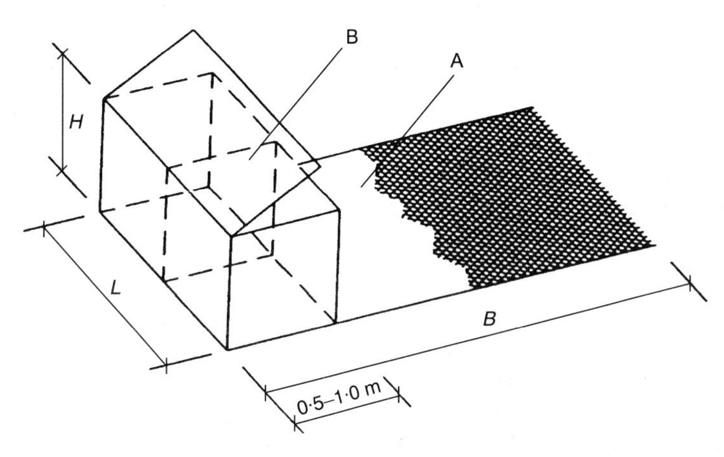

A – Patented Terramesh® unit of double twist hexagonal mesh type 8 × 10 zinc coated and PVC coated wire dia. 2·7 mm (3·7 mm o/d).

B – Diaphragm made with double twist hexagonal mesh type 8 × 10 zinc coated and PVC coated wire dia. 2·7 mm (3·7 mm o/d).

DESCRIPTION

System formed by double twisted hexagonal mesh with a front face consisting of stone-filled gabions (Terramesh) or soil (Reinforced Green Terramesh — water type). In both systems the front face and the soil reinforcement tail are made of a continuous mesh. Vertical or quasi-vertical walls with strong retaining capabilities can be achieved.

The galvanised steel mesh can be PVC coated. Vegetation can be encouraged to grow by part-filling the gabions with soil or by hydroseeding, among other methods.

CONSTRUCTION

Units are supplied folded and are opened in position. In the Terramesh system, the gabions are filled and lids are wired down before a geotextile is placed between the backfill and the gabions. In the Green Terramesh–water type, stones in-filled with soil are used at the face of the system and compacted soil at the back. This system is supplied with a geotextile fabric at the front mesh.

PRODUCT DATA

Terramesh® system
Mesh type 8 × 10
Zinc coated and PVC sleeved
Wire dia 2·7 mm (3·7 mm o/d)

Length *B*: m	Width *L*: m	Height *H*: m
3	2	0·50
4	2	0·50
5	2	0·50
3	2	1·00
4	2	1·00
5	2	1·00

Size tolerances ±5%

Reinforced Green Terramesh®
Mesh type 8 × 10
Zinc coated and PVC sleeved
Wire dia 2·7 mm (3·7 mm o/d)

Length *B*: m	Width *L*: m	Height *H*: m
3	2	0·45
4	2	0·45
5	2	0·45
3	2	0·60
4	2	0·60
5	2	0·60

See also above sketches. Size tolerances ±5%

Category Soil reinforcement systems/geomats
Product Fortrac

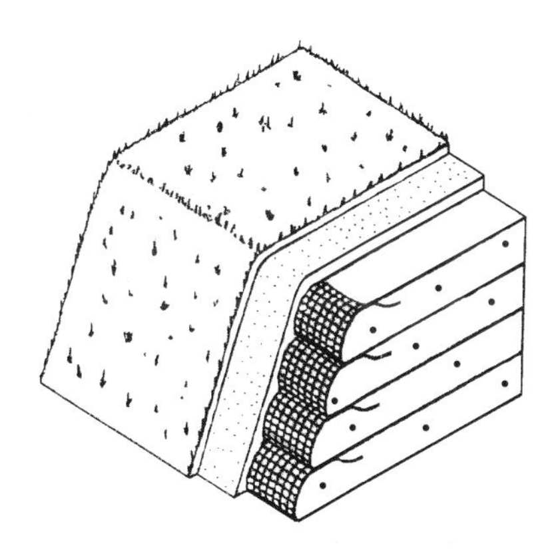

DESCRIPTION

Grid of high modulus polyester yarns with additional protective PVC layer used for reinforcement of slopes and for retaining walls. The product is available in a range of various tensile strengths, between 20 kN/m and 400 kN/m. Turf can be included to provide a green finish.

CONSTRUCTION

Fortrac geogrids are available in rolls with a standard width of 3·70 m (5 m widths can also be supplied) and 200 m length. The product is rolled out, cut to size and placed in layers typically at 500 mm centres and a certain minimum distance into the bank fill.

PRODUCT DATA

Refer to the manufacturer for detailed technical advice.

Category Soil reinforcement systems/geomats
Product Textomur

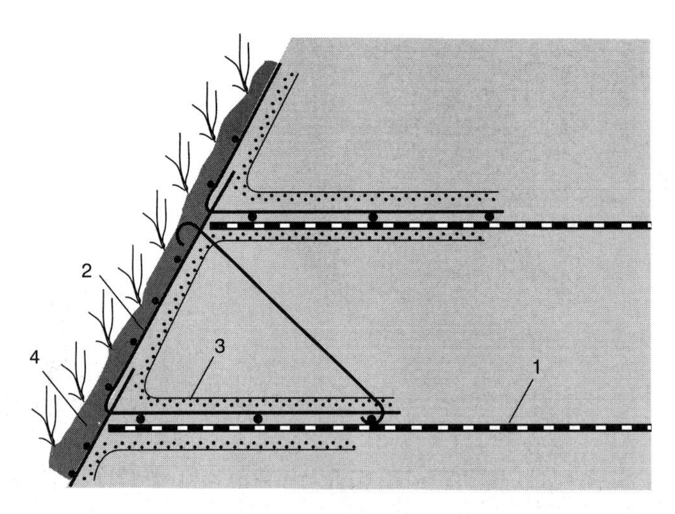

DESCRIPTION

Reinforcing geotextile with steel mesh framework used for retaining steep slopes (generally 60° angles). The face of the slope is hydraulically seeded to promote grass growth through the geotextile; shrubs can also be planted after grass establishment.

CONSTRUCTION

The installation is carried out in layers 500 mm thick. Each layer consists of a horizontally placed reinforcing geotextile with the steel mesh forming the front face, which is lined with a vegetation geotextile. The backfill is then placed and compacted. Hydroseeding and planting can then be carried out to provide vegetation on the front face.

PRODUCT DATA

Refer to the manufacturer for detailed technical advice.

Category Soil reinforcement systems/geomats
Product Enkamat

Enkamat *Enkamat A20*

Enkamat S *Enkazon*

DESCRIPTION

Three-dimensional random polyamide mattings ranging from a totally open structure to a fairly closed one at the back face. Vegetation is usually encouraged to grow through the mesh, which will enhance the erosion resistance of the system.

The range of products include: Enkamat S (formed by the open structure of Enkamat and a polyester reinforcement grid for use in very steep slopes); Enkamat A20 (filled with bitumen-bound stone chippings, used mainly in bank protection below water level for flow velocities below about 2·5 m/s); and Enkazon (turf pre-grown on Enkamat, mainly used on above water banks).

CONSTRUCTION

The surface should be well prepared to the required level. The mats should be anchored in trenches with pins at the top of the bank; anchoring should also be carried out at all edges and overlaps. In banks below water and up to 0·5 m above the water line, the mats should be filled with stone chippings (2–6 mm). Laying of mats should start from a downstream position to avoid upstream-facing overlaps.

PRODUCT DATA

Dimensions and weights

Enkamat

Type	Thickness: mm	Width: m	Length: m	Area/roll: m^2	Gross weight: kg
Open matting					
7010/1	10	1·00	150	150	40
7010/2	10	1·95	150	292·5	83
7010/4	10	3·85	150	577·5	165
7010/6	10	5·75	150	862·5	248
7020/1	20	1·00	100	100	41
7020/1	20	1·95	100	195	85
7020/4	20	3·85	100	385	169
7020/6	20	5·75	100	575	254
With flatback					
7220/1	18	1·00	60	60	25
7220/2	18	1·95	60	117	53
7220/4	18	3·85	60	231	105
7220/6	18	5·75	60	345	158

Enkamat S

Type	Tensile strength: kN/m	Thickness: mm	Width: m	Length: m	Area/roll: m^2	Gross weight: kg
S 20/1	20	15	1·00	100	100	48
S 20/3	20	15	2·90	100	290	149
S 35/1	35	16	1·00	100	100	70
S 35/3	35	16	2·90	100	290	217
S 55/1	55	17	1·00	100	100	78
S 55/3	55	17	2·90	100	290	241
S 110/1	110	18	1·00	50	50	58
S 110/3	110	18	2·90	50	145	181

Enkamat A20

Type	Thickness: mm	Weight: kg/m^2	Width: m	Length: m	Area/roll: m^2	Gross weight: kg
A 20/5	20	20	4·80	20	96	2400

*Enkazon**

Type	Enkamat type	Weight: kg/m^2	Width: m	Length: m
Enkazon 10	7010	23–35	1·00	5–20

*Enkazon is grown to order only

Enkamat material properties

Polymer type:	Polyamide 6	
Enkamat density:	approx. 25 kg/m^3	
Tensile strength (longitudinal) (DIN 53857):	7010:	> 1·5 kN/m
	7020:	> 2·3 kN/m
	7220:	> 1·8 kN/m
Strength at filament crossing points:	excellent because of total fusion of the filaments where they cross	
Ageing:	good resistance to weather influences and UV radiation because of stabilisation with carbon black and UV stabilisers	
Chemical resistance:	resistant to all chemicals in those concentrations which are normally concentrated in the earth and surface water	
Temperature resistance periods	from −30°C to + 100°C; can easily be installed during winter	
Inflammability:	low inflammability and low smoke formation; approved for use in tunnels	
Toxicity:	none; approved for use in drinking-water reservoirs; Enkamat is inert and not harmful to the environment	
Rodent attack:	no nutritive value; the tangled structure of the mat is unpleasant for burrowing animals and rodents	

Category Soil reinforcement systems/geomats
Product Tensar Mat

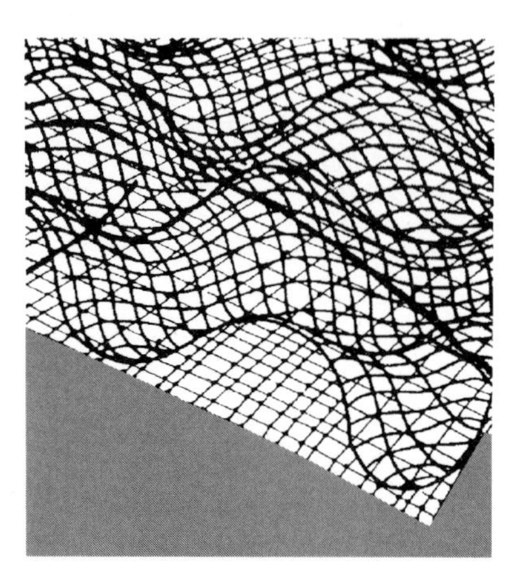

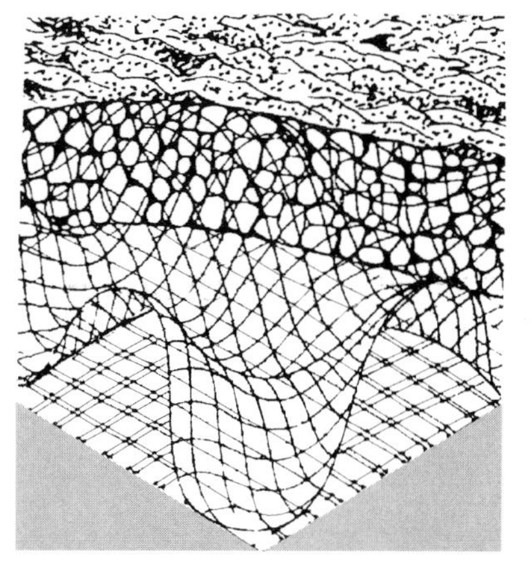

DESCRIPTION

Three-dimensional polyethylene mat with a cuspated surface and a flat base layer. The shape of the upper surface is such that pockets of soil are retained and vegetation is encouraged to grow in them and establish roots through the base layer.

The range of products include: Tensar Turf Mat (Tensar Mat incorporating pre-cultivated grass) for use when grass has no time to establish before possibility of erosion arises, and Macadamat (Tensar Mat incorporating bitumen-bound gravel) for use in permanently wetted areas, such as the lower reaches of embankments and river or channel beds.

CONSTRUCTION

Slopes should be prepared to required levels and 50–75 mm of topsoil can be placed on top of mat to enhance vegetation growth. The mat should be rolled down the slope and fixed with pegs at around 1–1·5 m centres onto the overlaps of adjacent mat strips. Trenches 450 mm wide by 250 mm deep should be used to bury the top and bottom ends of the mats. Grass can then be sown into the mat; soil, and sometimes fertilizer, is also brushed into the mat.

PRODUCT DATA

Quality control strength*: kN/m	Polymer	Weight: kg/m^2	Thickness: mm	Roll dimensions: m
3·2	Polyethylene	0·45	18	30 × 4·5 1·5 and 3·0 widths available to order

*Tested in accordance with BS6906 Part 1, following 24 hours test sample emersion in water at 20° ± 2°C.

Category Soil reinforcement systems/geomats
Product Geoweb

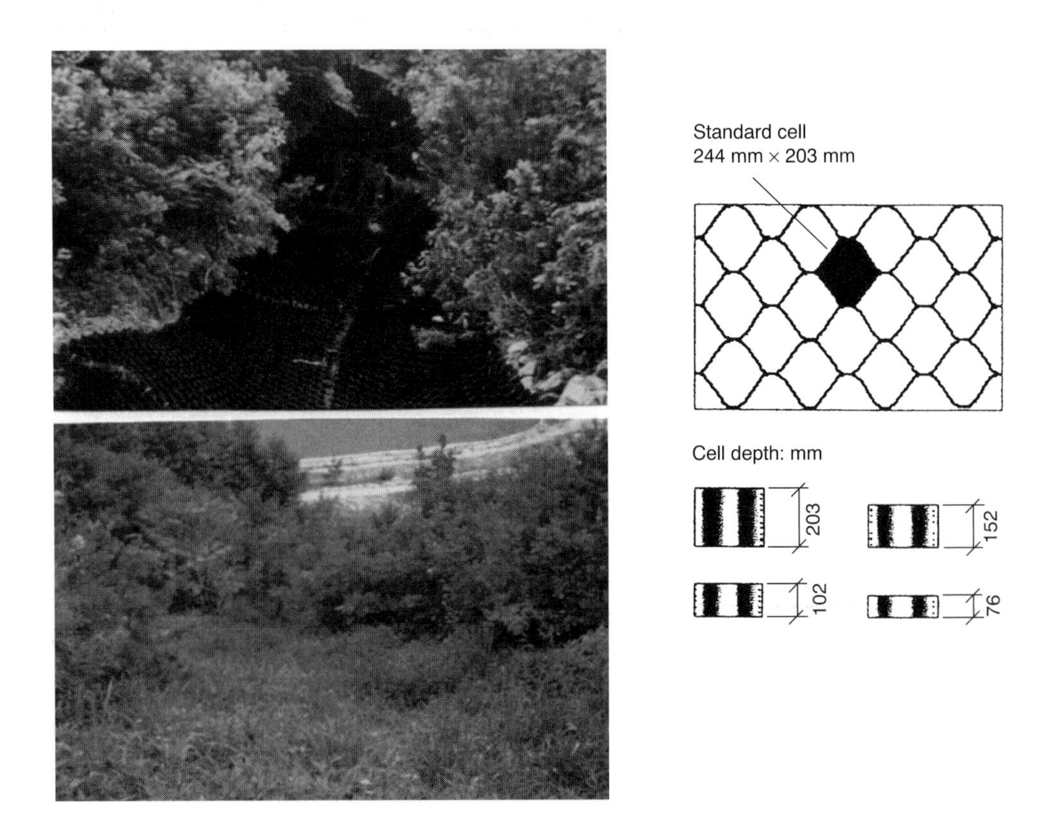

DESCRIPTION

Cellular confinement system in high density polyethylene, suitable for a variety of infill materials (e.g. topsoil with vegetation, gravels and sand, concrete or combinations of these). A range of cell wall types is available — smooth, textured (to increase friction and recommended for fine grain and concrete infills), and perforated (to increase friction with coarse aggregates, improve lateral drainage and promote root growth). Cell joints are ultrasonically welded and continuous polymeric tendons running through the cells can be used for anchorage purposes.

The system is available in various cell depths, sizes and colours and the manufacturer recommends the use of a nonwoven geotextile in most cases.

CONSTRUCTION

A suitable filter should be placed on graded and prepared banks and bed, on top of which the cellular system is laid in single or multiple layers. The system can be secured with a range of anchors to suit design and soil conditions; commonly used types are J-pin and earth anchors. Polyester or polyethylene-coated tendons through the cells can be used to anchor the system if required and can be fixed in trenches at the top of the bank.

Infills consisting of seeded topsoil are appropriate for upper slopes of channels, whereas concrete is used in areas of high flow velocities. Aggregate fills are used in a range of intermediate situations.

PRODUCT DATA

Comprehensive design services are available from manufacturer/supplier.

Category Soil reinforcement systems/geomats
Product Erosamat Type 3

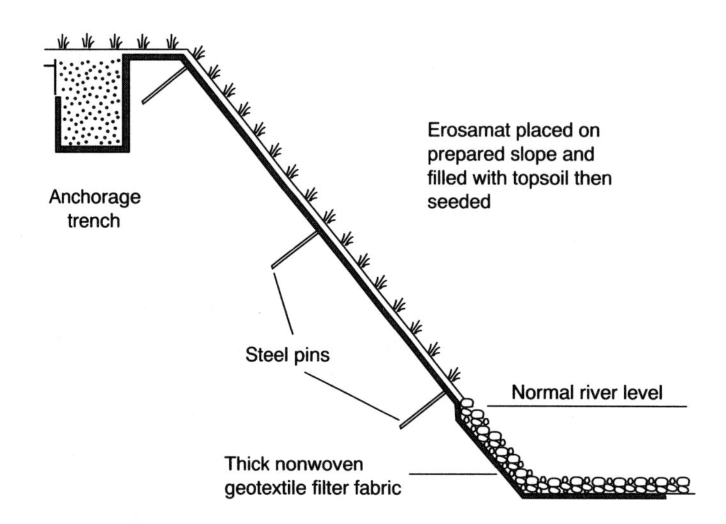

DESCRIPTION

Mattresses formed by high density polyethylene filaments which are thermally bonded in a random fashion. The mattresses are only 3 mm thick and relatively dense but are designed to allow vegetation growth through the openings. When used alone, this system is more suitable for erosion protection above mean water level; however, it can be used in conjunction with other products such as over cellular confinement systems.

CONSTRUCTION

Slopes should be prepared to the required grade and be free from obstructions or cavities. The mats should be buried at the edges into 300 mm wide by 250 mm deep trenches; additional transverse ditches are recommended in severe conditions to ensure adequate fixing of the mats. Adjacent mats should overlap by 100 mm and steel pins at 1 m spacings should be used to fix the mats down into the slope.

PRODUCT DATA

Polymer	High density polyethylene
Weight	850 g/m^2
Thickness	3 mm
Tensile	4 kN/m

Appendix 2. Specification of riprap for river applications

This appendix is intended to provide guidance on the specification of riprap to be used in the majority of situations that are likely to be encountered by river engineers. In most cases, the stable stone sizes found from application of the formulae given in Section 4.1 will be in the range 200–600 mm, which fall into the category of light gradings. Heavy and fine gradings will only seldom need to be specified as riprap cover layers. For protection works that require unusually big stone sizes, and for the design of very large extents of revetment in major rivers, the designer is recommended to consult other references, such as CIRIA/CUR (1991) — Appendices 1 and 2. Although produced for coastal and shoreline applications, this reference gives extensive guidance on specification and testing of quarried stone that is also appropriate for major river works and for the fine granular materials used as filters and underlayers.

It is recommended that the specification of riprap be comprised of the following items, which can be determined from standard tests (see CIRIA/CUR 1991). Depending on the complexity of the revetment work and on the availability of resources, the designer may decide not to include all these items in the specification; in this case it should be borne in mind that the serviceable life of the revetment may be shortened and more maintenance may be required.

A2.1. GRADINGS

More efficient and economic quarry productions, as well as improved quality control can be achieved with standardisation of riprap sizes. But, however practical and logical the idea of standard gradings is, it has not been widely implemented yet. At least in the UK, the designer should not expect to find quarries with a factory-like production of standard gradings ready to be picked. The CIRIA/CUR manual (1991) suggests a number of standard gradings within the light gradings category which, as mentioned above, are likely to include most of the stone sizes required for river bed and bank protection. These gradings are given either in terms of weight (kg) or dimensions (mm). Table A2.1 was reproduced from that manual, with some simplifications. In the table the grading classes are defined by two numbers: the lower class limit (LCL) and the upper class limit (UCL). The LCL is the size for which 10% of the stone is smaller and the UCL is the size for which 70% of the stone is greater. An alternative method for determining gradings is also presented in CIRIA/CUR (1991).

Table A2.1 Light grading class requirements (adapted from CIRIA/CUR, 1991)
Standard light gradings designated by *weight*

	Stone fraction (by weight: kg)			
Class: kg	< 2%	LCL 0–10%	UCL 70–100%	> 97%
10–60	2	10	60	120
60–300	30	60	300	450
10–200	2	10	200	300

Standard light gradings designated by *size*

	Stone fraction (by size: mm)			
Class: mm	< 2%	0–10%	70–100%	> 97%
200/350	100	200	350	400
350/550	250	350	550	650
200/500	100	200	500	550

Note: in this table the 10–60, 60–300 and 10–200 kg classes are approximately equivalent to the 200/350, 350/550 and 200/500 classes, respectively.

A2.2. SHAPE

In order to ensure that the stone is blockish in shape, it is important to specify that the ratio of the biggest to the smallest stone dimensions does not exceed a certain value. The minimum stone dimension is defined as the minimum distance between two parallel lines between which the stone can just be passed. The specification statement may be as follows:

> No more than 5% of stone in a sample shall have a ratio of maximum to minimum stone dimensions greater than 3°.

A2.3. DENSITY

Strength and durability are directly linked with stone density. Specification of this property may be given as:

> The mean density shall not be less than x kg/m^3, with 90% of stones having a density of at least (x – 100) kg/m^3. The value of x will normally be around 2600.

A2.4. WATER ABSORPTION

Water absorption is the single most important indicator of resistance to degradation and a good indicator of weathering resistance. Requirements on water absorption will depend on the quality of the rock needed for the revetment. For good quality rock, a typical example of specification will state that:

The average water absorption must be less than 2% and the water absorption of nine of the individual stones less than 2·5%.

A2.5. RESISTANCE TO WEATHERING

Resistance to weathering can be determined from some other properties in addition to water absorption. It should be determined by specifying:

- magnesium sulphate soundness
- average freeze–thaw weight loss below 0·5% for good quality rock
- non-existence of deleterious secondary minerals.

A2.6. RESISTANCE TO ABRASION

This can be an important aspect, particularly where riprap is subjected to flows heavily loaded with sediment, such as in gravel upland rivers or in very turbulent flows. The mill abrasion resistance index is the standard indicator that should be specified in those cases.

A2.7. OTHER ASPECTS

Other properties that the designer may want to specify are:

- stone colour
- stone integrity
- resistance to impact
- absence of impurities in the stone.

It is also advisable to include in the specification document clauses concerning quality control of the materials at the quarry, on site during construction, and after construction during the monitoring and inspection stages.

Appendix 3.
Specification of gabion mattresses

Major manufacturers of gabion mattresses are usually able to supply custom-made products to meet individual requirements if the need arises. These individual requirements may be driven by durability, performance or economic factors, but some basic aspects, which are summarised below, should always be considered in the specification of gabion mattresses.

A3.1. MESH

Type of mesh. This makes the distinction between welded and woven material.

Size of mesh opening. This size should be such that the estimated stable stone will be contained by the mesh.

Mesh wire. Specification should be made of the material of the mesh (galvanised steel, PVC coated wire or polyethylene) and of the diameter of the wire. The materials should conform to relevant standards (for example, in the UK steel wire should conform to British Standard BS1052:1980, 1986).

Protective coating. Similarly to the mesh material, coatings should conform to relevant standards (for example to British Standard BS443: 1982, 1990, which regulates galvanisation).

A3.2. MATTRESS SIZE

The mattress size can be given in terms of length, width and thickness. However, it is usually sufficient to specify mattress thickness since this (and the stone size) defines the protection characteristics of the mattress.

A3.3. REINFORCEMENT OF MATTRESS EDGES AND JOINTS

It is advisable to specify some reinforcement of corners and jointing by means of continuous lacing or rings with the same specification as the wire mesh.

A3.4. STONE FILL

Specification of size is normally in terms of D_{50} but the smaller stone fraction should be such that loss of particles through the mesh holes will not occur. Specification may also include type and shape of stone.

A3.5. ASSEMBLING AND PLACING OF MATTRESSES

Assembling involves fitting and fixing of lids (see A3.3 on reinforcement of edges). Placing and fixing at edges should be carried out according to detail drawings.

Appendix 4. Specification of concrete block mattresses

This appendix lists the major items that should be included in the specification of river revetments formed by pre-cast concrete blocks. The guidance is given in general terms and details will depend on the particular site requirements and conditions.

A4.1. BLOCK DETAILS

- Minimum thickness (usually in mm).
- Minimum weight per unit of area (in kg/m^2).
- Unit weight (in kg). Both the maximum and minimum block weights may need to be specified: the maximum weight when hand placing is dictated by the form of installation, and the minimum weight by the need to comply with stability requirements.
- Open area of block. Usually given in percentage and defined as the area of voids divided by the block surface area, the open area is important where adequate relief of pore pressure and/or vegetation cover are required.
- Surface finish. Specification may require smooth concrete finish complying with the relevant standard or other types such as rock finish.

A4.2. CONCRETE QUALITY

In the UK, concrete quality standards are specified by British Standards and European norms. At the time of production of this book, concrete blocks are manufactured according to the regulations for concrete given in BS5328 (1990, 1991), for cement in BS12:1996, and for aggregates in BS882:1983.

- Concrete grade — common grades are C50 and C35, as specified in BS5328. Alternatively, concrete strength can be specified in terms of N/mm^2.
- Cement — usually Portland cement complying with BS12.
- Maximum aggregate size — aggregates should comply with BS882, with maximum aggregate size of 10 mm.
- Water/cement ratio — usually limited to 0·45 or 0·50.
- Resistance to sulphate attack — depending on type of exposure, in most situations blocks should be Class 2 or 3 of Table 7 in BS5328.

A4.3. SYSTEM DETAILS

The specification of the block mattress depends, to a large extent, on whether the system is formed by loose or by cabled blocks. Cabled mattresses and geotextile-bond blocks will respectively require specification of the cable material and ways of fixing, or of the geotextile membrane and type of bonding.

- Width and length of panels — these should suit areas shown on layout drawings. When the panels are assembled off-site there may be limitations as to the maximum width of panels that can easily be transported.
- Block interaction and pattern layout — when interlocking of blocks is a requirement, this should be specified, as well as the pattern to which the blocks are to be laid.
- Cables (for cabled mattresses only) — the specification should mention the cable material (usually one of the following: galvanised wire, stainless steel and synthetic fibres) and the direction(s) of cabling (one or two directions).
- Assembling and placement:
    - site or works assembled
    - changes in direction and construction at edges. It should be specified that all cut blocks in cabled mattresses should be at least half full block size and retained by at least one cable
    - jointing at edges — should be carried out according to detail drawings and manufacturers' recommendations
    - method of fixing cables.
- For geotextile-bond mattresses only:
    - properties of geotextile
    - bonding characteristics.

Appendix 5. Specification of geotextiles

The specification of geotextiles should have regard to the following properties.

- *Engineering function.* In river engineering this will principally be a filtering function, where permeability and retention of fine solids are the major factors. These are defined in terms of the O_{90} (the opening size of the geotextile corresponding to the diameter of the largest particles able to pass through it), of the permeability (m/s), and also of the *type of geotextile* to use (woven or nonwoven, material, thickness).
- *Durability during construction.* It may be important to specify this property to avoid damage, particularly when large angular riprap is to be dumped onto the geotextile. Specification should be made of the *ultimate tensile strength* (kN/m width) and of *elongation at break* (% strain at ultimate load). Also important may be the capacity to withstand exposure to ultra-violet radiation during installation.
- *Long-term durability.* This is determined by the site conditions and defined by the *type of material* (for example, polyester should not be specified if placed next to setting cement or concrete, or in situations of water or ground pH above 10).

One effective way of specifying geotextiles is to arrange the required properties into the following categories, as suggested by Rankilor (1997):

1. imperative (those properties that all bidders need to fulfil, such as the O_{90}, delivery dates, type of geotextile
2. important (those properties that will allow distinction between bidders, such as, for example, the colour of geotextile or the maximum width that can be supplied) and
3. information (a typical example is the weight of the geotextile, which does not relate directly to its engineering properties but can be an important constraint during installation).

Appendix 6.
Examples of edge details

It is often observed that much of the damage that occurs in revetment systems tends to be found at the edges, which are generally weak points in a protection scheme. This can be due to either inadequate design or faulty construction and can, in many cases, lead to the collapse of the whole system. Therefore the importance of careful detailing at edges and transitions should not be underestimated.

There are three areas that require special attention: the toe of banks, junctions with structures or other types of revetment and, perhaps less important in river protection schemes, the crest of embankments. In all these situations, the changes in the boundaries affect the direction of the flow and increase the hydraulic forces exerted on the revetments.

A6.1. TOE DETAILS

Since many types of failure involve the undermining of the toe due to insufficient support, adequate protection of the toe is essential to ensure stability of the bank. In order to design stable toe protection it is necessary first to estimate the anticipated scour depth below the existing bed level (see Section 2.4.3). Once this is determined, the protection should be designed to extend to about 50% beyond that level or to reach bedrock if possible.

The two most efficient forms of toe protection are the provision of cut-offs (vertical walls that obstruct the erosive process, avoiding its spread towards the bank) and of falling or launching aprons (formed by riprap designed to follow the contours of the scour hole as it develops). They also increase the seepage path therefore reducing the risk of loss of fines from the bank soil. Cut-offs formed by piling walls are illustrated in Figure 4.25 and falling aprons can be seen in Figure 2.5. Other methods, such as the reinforcement of the toe by dumping extra stone or rock rolls, or burying block stone or gabions at a level below that of anticipated scour, are commonly used and can also provide adequate protection in situations of less severe erosive potential. Some of the many types of toe protection possible are illustrated in Figure A6.1, where d_s is the expected scour depth.

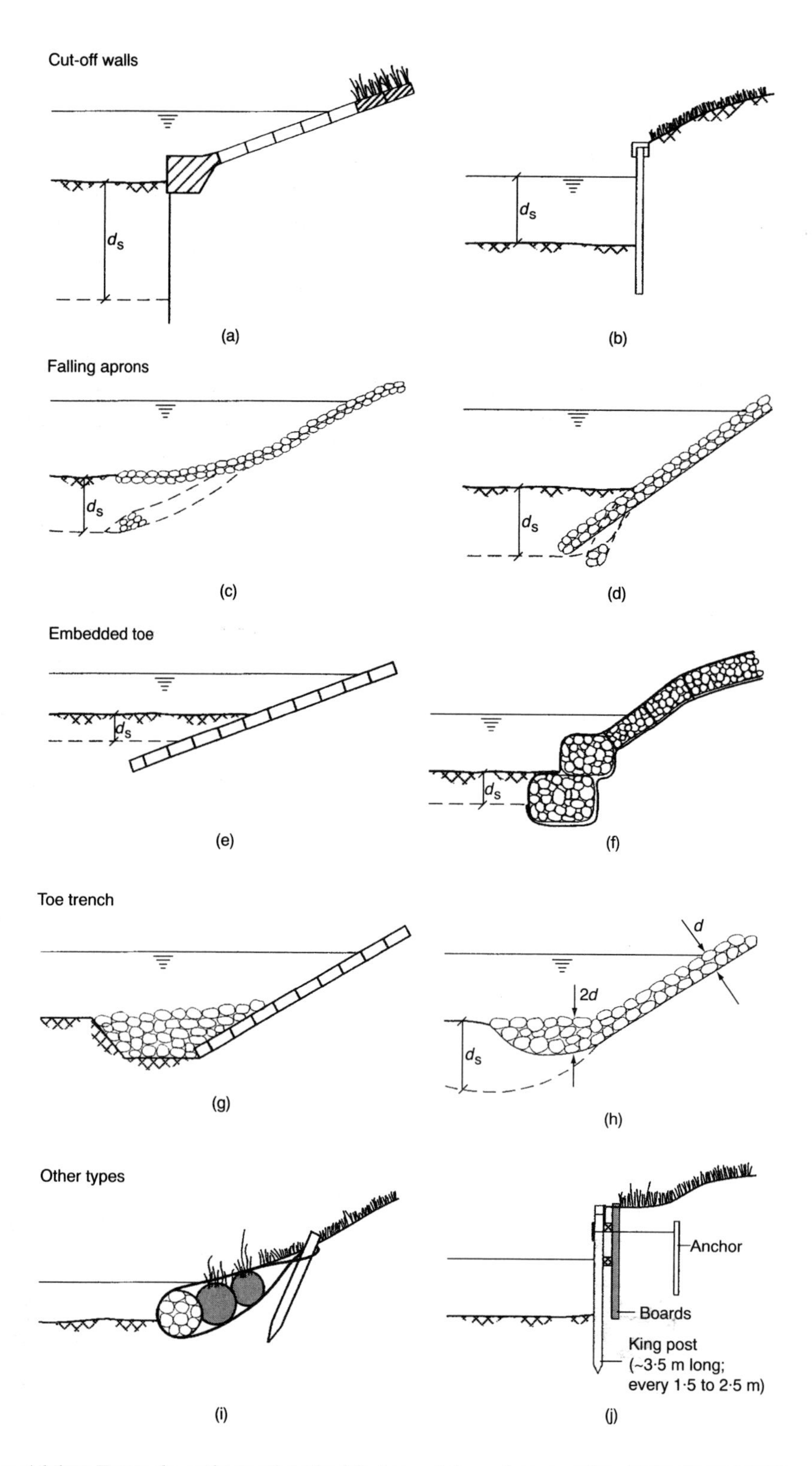

Figure A6.1. Examples of toe details (d_s is anticipated scour depth): Cut-off walls (a) with concrete toe beam; (b) sheet piling; Falling aprons (c) resting on existing bed level; (d) embedded apron; Embedded toe (e) embedded concrete block mat; (f) embedded gabions; Toe trench (g) loose stone protecting base of a concrete mat; (h) toe trench for riprap revetment; Other types (i) rock and planted rolls; (j) timber piling

A6.2. JUNCTIONS AND TRANSITIONS

The general rule about revetment edges and junctions is that the protection should be extended beyond the area experiencing erosion. The extent of protection varies according to the situation and should, whenever possible, be determined using survey data that include vertical velocity profiles and turbulence levels (see Section 2.4.1). Very approximate rules can be given to determine the minimum extent of protection:

- downstream of culverts, protection should be extended to at least 2·5 times the flow depth and to 4 times the depth in highly turbulent environments
- the protection of outer bends should be extended to 1·5 times the mean water surface width beyond the eroded reach and to a minimum of 1 water width upstream
- around bridge piers the minimum recommended extent of protection should be approximately 1·5 times the pier width.

Transitions between protected banks and vertical or sloping fixed structures can be achieved in a number of different ways, as shown in Figure A6.2. It is necessary in these situations to use revetment types that have sufficient flexibility to follow the required soil contours in order to produce a smooth transition. Examples of these types include riprap, loose concrete blocks and gabions. It is also good practice to choose revetments with hydraulic roughnesses that are intermediate between that of the main revetment and that of the structure or

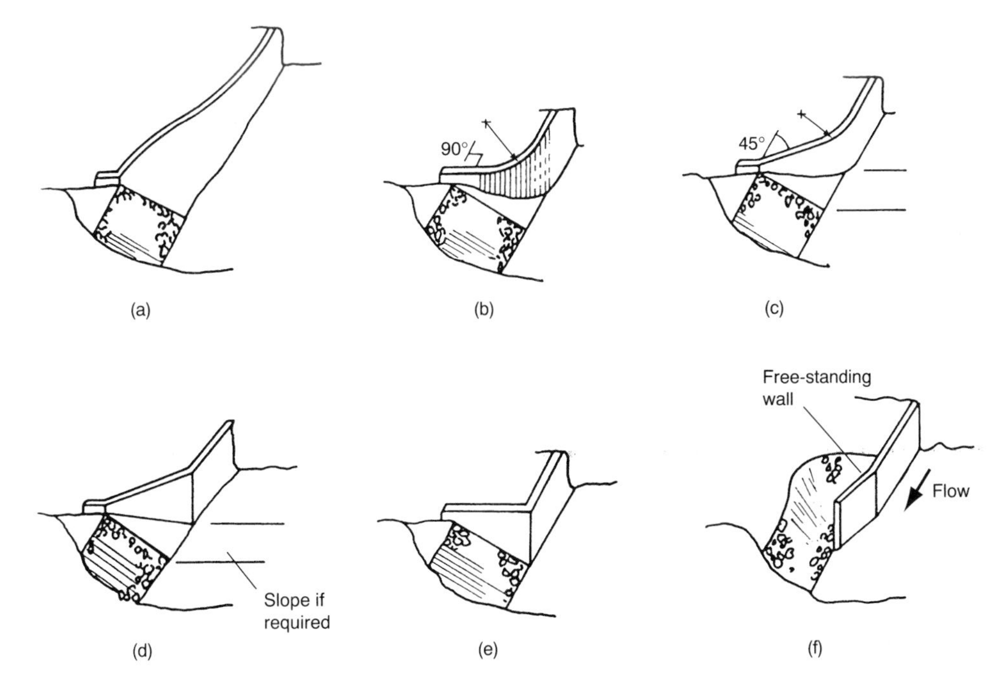

Figure A6.2. Examples of transitions between structures and protected banks (from Hemphill and Bramley, 1989): (a) warped (gradual); (b) cylinder-quadrant; (c) broken-back (with curvature); (d) broken-back; (e) square-ended; (f) 'onion' outlet

natural bank. This can sometimes be achieved by locally grouting the revetment, which has the added advantage of also increasing its stability in these particularly vulnerable areas. The edges of revetment mats placed on slopes should also be turned into the bank and buried to ensure stability.

A6.3. CREST

One of the aspects that needs to be addressed is the elevation of the bank at which protection should be terminated. In many cases this is dictated by the presence of waves and therefore an estimation needs to be made of the likely extent of '. This depends on the roughness of the revetment and on the nature of the waves (e.g. plunging or breaking waves) but the following simple rules of thumb can be used as first approximations to determine the height above still water level h of the main bank protection:

$h = 2H$ for riprap

$h = 4H$ for concrete mattresses or asphalt mats

where H is the design wave height. See also Section 2.3.2 for more information on wave attack.

Some form of crest protection is usually also required in flood protection schemes where severe damage to flood banks during flood events is not acceptable. However, limited damage that will not jeopardise the integrity of the scheme may be allowed in many cases. Therefore the type of protection used in the upper banks can be designed for a lower degree of stability than that of the main channel.

References

Bathurst, J. C. (1979). Distribution of boundary shear stress in rivers. Proc. 10th Annual Geomorphology Symp., Binghampton, New York, 21–22 September 1979, ed. D. D. Rhodes and G. P. Williams.

Berry, P. L. and Reid, D. (1987). *An introduction to soil mechanics.* McGraw-Hill, New York.

Breteler, M. K., Pilarczyk, K. W. and Smith, G. M. (1995). *Geotextiles in bed and bank protection structures.* Publication No. 488, Delft Hydraulics, Netherlands, February 1995.

Breusers, H. N. C. and Raudkivi, A. J. (1991). Scouring, *Hydraulic structures design manual,* IAHR, Balkema, Rotterdam. ISBN 90 6191 983 5.

British Steel (1997). *Piling handbook.* 7th ed. British Steel, Image Colour Print Ltd, East Yorkshire, UK. ISBN 0 95299 1209.

BSI (1986). Specification for mild steel wire for general engineering purposes, *British Standard BS1052: 1980.* British Standards Institution. ISBN 0 580 11323 X.

BSI (1990). Specification for testing zinc coatings on steel wire and for quality requirements, *British Standard BS443: 1982.* British Standards Institution. ISBN 0 580 12612 9.

BSI (1990, 1991). Concrete, *British Standard BS5328: Parts 1 to 4: 1990, 1991.* British Standards Institution. ISBNs: 0 580 2026 7 4; 0 580 20274 7; 0 580 18979 1; 0 580 18980 5.

BSI (1992). British Standard Specification for aggregates from natural sources from concrete, *British Standard BS882: 1992.* British Standards Institution. ISBN 0 580 21 463 X.

BSI (1994). Code of practice for earth retaining structures, *British Standard BS8002: 1994.* British Standards Institution. ISBN 0 580 228 26 6.

BSI (1996). Specification for Portland cement, *British Standard BS12: 1996.* British Standards Institution. ISBN 0 580 25343 0.

Capper, P. L. and Cassie, W. F. (1969). *The mechanics of engineering soils.* E & FN Spon, London.

Chow, V. T. (1973). *Open-channel hydraulics.* McGraw-Hill, New York. ISBN 0 07 Y85906 X.

CIRIA (1976). *Design of riprap slope protection against wind wave attack.* D. M. Thompson and R. M. Shuttler. Construction Industry Research and Information Association, Report 61.

CIRIA (1987). *Design of reinforced grass waterways.* H. W. M. Hewlett *et al.* Construction Industry Research and Information Association, Report 116.

CIRIA/CUR (1991). *Manual on the use of rock in coastal and shoreline engineering.* Construction Industry Research and Information Association, Special publication 83. Centre for Civil Engineering Research and Codes, Report 154. CIRIA, London and CUR, Netherlands.

Craig, R. F. (1987). *Soil mechanics.* Van Nostrand Reinhold.

CUR (1995). *Manual on the use of rock in hydraulic engineering.* Report 169. Balkema, Rotterdam. ISBN 90 5410 6050.

EAU 1990 (1992). *Recommendations of the committee for waterfront structures; harbours and waterways.* Ernst & Sohn, Berlin. ISBN 3 433 01237 7.

Escarameia, M. and May, R. W. P. (1992). *Channel protection downstream of structures.* Report SR 313, April 1992, HR Wallingford.

Escarameia, M., May, R. W. P. and Atkins, R. (1995). *Field measurements of turbulence in rivers.* Report SR 424, April 1995, HR Wallingford.

Escarameia, M. (1995). *Channel protection; gabion mattresses and concrete blocks.* Report SR 427, July 1995, HR Wallingford.

Hedges, T. S. (1990). The hydraulic climate. Design workshop *Geogrids and geotextiles in the maritime and waterways environment,* 26–27 September, Liverpool, 1990.

Hemphill, R. W. and Bramley, M. E. (1989). *Protection of river and canal banks.* CIRIA (Construction Industry Research and Information Association), Butterworths, London.

Henderson, F. M. (1966). *Open channel flow.* Macmillan Publishing Co. Inc, New York.

Knight, D. W. and Shiono, K. (1996). *Floodplain processes.* Eds M. G. Anderson, D. E. Walling and P. D. Bates. Wiley & Sons, Chichester, UK.

Knight, D. W., Yuen, K. W. H. and Al-Hamid, A. A. I. (1994). *Mixing and transport in the environment*. Eds K. J. Beven, P. C. Chatwin and J. H. Millbank. Wiley & Sons, Chichester, UK.

May, R. W. P. and Willoughby, I. R. (1990). *Local scour around large obstructions*. Report SR 240, HR Wallingford.

Maynord, S. T. (1993). Corps riprap design guidance for channel protection. Preprints of the *Int. Riprap Workshop. Theory, policy and practice of erosion control using riprap, armour stone and rubble*. Fort Collins, Colorado, USA.

McConnell, K. J. (1998). *Revetment systems against wave attack. A design manual*. Thomas Telford Publishing, London.

Moffat, A. I. B. (1990). Introduction to soil mechanics. Lecture notes. University of Newcastle upon Tyne.

Morgan, R. P. C., Collins, A. J., Hann, M. J., Morris, J., Dunderdale, J. A. L. and Gowing, D. J. G. (1998). *Waterway bank protection: a guide to erosion assessment and management*. Environment Agency, Bristol, UK.

Neill, C. R., ed, (1973). Guide to bridge hydraulics. Road and Transport Association of Canada, University of Toronto Press. ISBN 0 8020 1961 7.

Nezu, I. and Nakagawa, H. (1993). *Turbulence in open-channel flows*. IAHR monograph. Balkema, Rotterdam. ISBN 90 5410 1180.

NRA (undated). Understanding riverbank erosion from a conservation perspective. National Rivers Authority, UK.

Perkins, J. A. (1994). Practical aspects of river bank protection in the developing world. *IWEM Seminar on river erosion and its prevention*. HR Wallingford, 11 March 1994.

PIANC (1987). *Guidelines for the design and construction of flexible revetments incorporating geotextiles for inland waterways*. Report of working Group 4 of the Permanent Technical Committee I, Supplement to Bulletin No. 57, Belgium. ISBN 2 87223 000 9.

Pilarczyk, K.W. (1990). Stability criteria for revetments. *Proc. of the 1990 National Conf. on Hydraulic Engng, Am. Soc. Civ. Engrs*. Eds H. H. Chang and J. C. Hill, San Diego, USA.

Przedwojski, B., Btazejewski, R. and Pilarczyk, K. W. (1995). *River training techniques. Fundamentals, design and applications*. Balkema, Rotterdam. ISBN 90 5410 1962.

Rankilor, P. R. (1981). *Membranes in ground engineering*. Wiley & Sons, Chichester, UK. ISBN 0 471 278084.

Rankilor, P. R. (1994). *UTF geosynthetics manual*. UCO Technical Fabrics NV, Belgium.

Rankilor, P. R. (1997). *Geosynthetics design notes*. Course on Practical Application of Geotextiles to Civil Engineering Problems, Ascot, UK, July 1997.

Raymond, G. P. and Giroud, J. P. (eds) (1993). *Geosynthetics case histories*. Int. Society for Soil Mechanics and Foundation Engineering.

Richards, K. (1982) *Rivers-form and process in alluvial channels*. Methuen & Co Ltd, UK and USA. ISBN 0 416 74900 3.

Saville, T., McClendon, E. W. and Cochran, A. L. (1962). Freeboard allowances for waves in inland reservoirs. *J. Waterways and Harbors Div., Am. Soc. Civ. Engrs*, **88**, WW2, 93–124.

Tennekes, H. and Lumley, J. L. (1972). *A first course in turbulence*. The MIT Press, Cambridge, USA.

Terzaghi, K. and Peck, R. B. (1948). *Soil mechanics in engineering practice*. Wiley & Sons, New York.

The Institution of Civil Engineers (1996). *Specification for piling and embedded retaining walls*. Thomas Telford Publishing, London. ISBN 0 7277 2566 1.

Thorne, C. R., Abt, S. R., Barends, F. B. J., Maynord, S. T. and Pilarczyk, K. W. (eds) (1995). *River, coastal and shoreline protection – erosion control using riprap and armourstone*. Wiley & Sons, Chichester, UK. ISBN 0 471 94235 9.

Thorne, C. R., Reed, S. and Doornkamp, J. C. (1996). *A procedure for assessing river bank erosion problems and solutions*. R&D Report 28, National Rivers Authority. ISBN 1 873160 31 3.

US Army Corps of Engineers (1981). *Final report to Congress. The Streambank Erosion Control Evaluation and Demonstration Act of 1974*.

van der Meer, J. W. (1988). Deterministic and probabilistic design of breakwater armour layers. *J. Waterway, Port, Coastal and Ocean Engng, Am. Soc. Civ. Engrs*, **114**, WW1, 66–80

Yarde, A. J., Banyard, L. S. and Allsop, N. W. H. (1996). *Reservoir dams: wave conditions, wave overtopping and slab protection*. Report SR 459, April 1996, HR Wallingford.

Yuen, C. and Fraser, D. (1979). *Digital spectral analysis*. Pitman Publishing Limited, London.

Index

RECEIVED
FEB - 8 1999
ENGINEERING LIBRARY